David Lloyd (Ed.)

Flow Cytometry in Microbiology

With 74 Figures

Springer-Verlag
London Berlin Heidelberg New York
Paris Tokyo Hong Kong
Barcelona Budapest

David Lloyd, BSc, PhD, DSc
Professor of Microbiology, University of Wales College of
Cardiff, PO Box 915, Cardiff CF1 3TL, Wales

Cover illustration: bis-(1,3-dibutybarbituric acid)-trimethine oxanol stained
Eschericha coli NCTC 10418 (unpublished experiment of David Mason and David
Lloyd) showing separation of viable and non-viable sub-populations. *x* axis:
Forward Light Scatter. *y* axis: Fluorescence.

*QR
69
.F56
1993*

ISBN 3-540-19796-6 Springer-Verlag Berlin Heidelberg New York
ISBN 0-387-19796-6 Springer-Verlag New York Berlin Heidelberg

British Library Cataloguing in Publication Data
Flow Cytometry in Microbiology
  I. Lloyd, David
  576
  ISBN 3-540-19796-6

Library of Congress Cataloging-in-Publication Data
Flow cytometry in microbiology / David Lloyd, ed.
    p.   cm.
  Includes bibliographical references and index.
  ISBN 0–387–19796–6: $120.00 (est.). — ISBN 3–540–19796–6
  1. Flow cytometry.   2. Microbiology—Technique.   I. Lloyd, David,
1940–
QR69.F56F56   1993   92–38958
576′.028—dc20   CIP

Typeset by Best-set Typesetter Ltd., Hong Kong.
12/3830-543210 Printed on acid-free paper

Preface

As yet, flow cytometry is not used so widely in microbiology as in some other disciplines. This volume presents contributions from research microbiologists who use flow cytometry to study a diverse set of problems. It illustrates the power of the technique, and may persuade others of its usefulness. Most of the contributors gathered in Cardiff on 23 October 1991, at a meeting organized for the Royal Microscopical Society by Dr. Richard Allman, but the content of their chapters is not limited by the discourse of that meeting, and for balance other experts were invited to write for this book. *Flow Cytometry in Microbiology* thus represents the first collection of articles specifically devoted to the applications of a technique which promises so much to those investigating the microbial world.

Cardiff, 1992   David Lloyd

Contents

Contributors

Richard Allman
Microbiology Group, PABIO, University of Wales College of Cardiff, PO Box 915, Cardiff CF1 3TL, Wales

Lynne Boddy
Microbiology Group, PABIO, University of Wales College of Cardiff, PO Box 915, Cardiff CF1 3TL, Wales

Erik Boye
Department of Biophysics, Institute for Cancer Research, Montebello 0310, Oslo 3, Norway

Michael Brailsford
Chemunex, The Jeffreys Building, St. John's Innovation Park, Cowley Road, Cambridge CB4 1UF, England

Peter H. Burkill
Plymouth Marine Laboratory, Prospect Place, Plymouth PL1 3DH, England

Elizabeth A. Carter
SmithKline Beecham Pharmaceuticals, Great Burgh, Yew Tree Bottom Road, Epsom, Surrey KT18 5XQ, England

Kenneth Coleman
SmithKline Beecham Pharmaceuticals, Brockham Park Research Centre, Betchworth, Surrey RH3 7AJ, England

Alex Cunningham
Department of Physics and Applied Physics, Strathclyde University, John Anderson Building, 107 Rottenrow, Glasgow G4 ONG, Scotland

Chris L. Davey
Department of Biological Sciences, University College of Wales,
Aberystwyth, Dyfed SY23 3DA, Wales

Hazel M. Davey
Department of Biological Sciences, University College of Wales,
Aberystwyth, Dyfed SY23 3DA, Wales

Julian P. Diaper
Department of Genetics and Microbiology, Life Sciences
Building, PO Box 147, Liverpool L69 3BX, England

Jerome Durodie
SmithKline Beecham Pharmaceuticals, Brockham Park Research
Centre, Betchworth, Surrey RH3 7AJ, England

Clive Edwards
Department of Genetics and Microbiology, Life Sciences
Building, PO Box 147, Liverpool L69 3BX, England

Stephen Gatley
Chemunex SA, 41 Rue de 11 Novembre 94700, Maisons Alfort,
France

Pamela A. Hunter
SmithKline Beecham Pharmaceuticals, Great Burgh, Yew Tree
Bottom Road, Epsom, Surrey KT18 5XQ, England

Douglas B. Kell
Department of Biological Sciences, University College of Wales,
Aberystwyth, Dyfed SY23 3DA, Wales

Arseny S. Kaprelyants
Bakh Institute of Biochemistry, Russian Academy of Sciences,
Leninskii Prospekt 33, 117071, Moscow, Russia

David Lloyd
Microbiology Group, PABIO, University of Wales College of
Cardiff, PO Box 915, Cardiff CF1 3TL, Wales

Richard Manchee
Chemical Defence Establishment, Porton Down, Salisbury,
Wiltshire SP4 OJQ, England

David Mason
Microbiology Group, PABIO, University of Wales College of
Cardiff, PO Box 915, Cardiff CF1 3TL, Wales

Colin W. Morris
Department of Computer Studies, University of Glamorgan,
Pontypridd CF37 1DL, Wales

Frank E. Paul
SmithKline Beecham Pharmaceuticals, Great Burgh, Yew Tree
Bottom Road, Epsom, Surrey KT18 5XQ, England

Roger Pickup
Institute of Freshwater Ecology, Windermere Laboratory, Far
Sawrey, Ambleside, Cumbria LA22 OLP, England

Jonathan Porter
Department of Genetics and Microbiology, Life Sciences
Building, PO Box 147, Liverpool L69 3BX, England

Harald B. Steen
Department of Biophysics, Institute for Cancer Research, The
Norwegian Radium Hospital, Montebello O310, Oslo, Norway

Glen A. Tarran
Plymouth Marine Laboratory, Prospect Place, Plymouth PL1
3DH, England

Michael J. Wilkinson
SmithKline Beecham Pharmaceuticals, Brockham Park Research
Centre, Betchworth, Surrey RH3 7AJ, England

Chapter 1

Flow Cytometry: A Technique Waiting for Microbiologists

David Lloyd

Introduction

The world of microorganisms provides a playground both for those fascinated by the prospects of understanding the nature of life, and for those interested in solving everyday problems arising from microbial activities. Both fundamental studies and applied aspects of microbiology have become "illuminated" in recent years by the techniques of flow cytometry (Boye and Løbner-Olesen 1990). It is, however, somewhat disappointing, considering the truly remarkable demonstrations of the usefulness and power of these approaches, that only a rather slow adoption has ensued. The reasons for this are not hard to find, and do not in any way detract from the great potential of flow cytometry as amply illustrated by existing literature. Cost may become less of a constraint as suitable low-power lasers become available (Shapiro 1985). It is also worth remembering that a high-pressure mercury arc lamp can give results as good as (or better than) a 5 W argon-ion laser in many applications (Peters 1979).

Every microbiologist, irrespective of his or her interests, uses a light microscope. For those investigating biochemical and molecular aspects, it may be only to check the "purity" of cultures, to count organisms, to discriminate between live and dead individuals, to differentiate intact from broken organisms, to observe the progress of sphaeroplast formation, or to check distributions of the larger organelles between subcellular fractions. For microbial ecologists, a great diversity of organisms is always present in the system of study, and an added layer of complexity is provided by spatial heterogeneities. Thus in most natural environments, organisms grow as biofilms rather than freely suspended in water. Light microscopy and all its latest technical advances (digital imaging, confocal scanning, etc.) provides the basis for studies of interacting organisms in mixed populations essential to our deeper insights into nutrient cycling processes in natural ecosystems. In those industrial processes which rely on the controlled or optimized activities of microbes, reproducible productivity depends on maintenance of the microbial population in a genetically homogeneous state, and progress requires strain improvement programmes. Here again, in the management of all microbial fermentations, microscopy plays a key role. The rigorous exclusion of contaminants and the rapid diagnosis of the presence of intruders become of paramount importance. Sometimes less immediately obvious, but in the long-term equally expensive problems arise from microbial contamination and spoilage. These afflict a wide range of industrial processes from steel-making to milk and food production. Microbes spoil paintwork, dartboards and ships' engines, and corrode the submerged steel of oil rigs (Hill and Genner 1981). Microscopy enables identification of the culprits, and monitoring of the progress of treatment. The detective work in these fields is not too different from that required of the medical microbiologist.

Microbiologists thus face similar problems irrespective of their spheres of activity. These are stated simply as follows: (1) to determine the diversity of organisms present, (2) to distinguish minority (contaminating) species, (3) to determine growth rates, and (4) to check whether treatment has been effective.

Quantitative information is hard won from the more conventional microscopical techniques; for example, the standard deviation (SD) for a count of n organisms is given by $\pm 100/\sqrt{n}(\%)$, or $\pm 1\%$ for 10 000 counts. Even for the most skilled microscopist, acquisition of this information requires a whole morning's work; it would perhaps be unreasonable to expect a repeat in the afternoon! Differential counting is even more slowly achieved, and even more demanding. Digital imaging procedures can help, but imprecision may arise in automated systems where distributions of organisms show overlaps, or when they cannot be easily distinguished from inorganic particles.

Almost everything that can be done using a microscope can be done more quantitatively and more quickly with a flow cytometer. The prospect of examining 1000 organisms per second is encouragement enough. Moreover most of the information available from flow cytometry cannot be obtained by any other technique.

The Heterogeneity of a Pure Culture

Whenever information is obtained from a population of organisms, it cannot be assumed that each and every individual contributes equally to the total. That this is so, comes about from several possible sources – almost always ignored by investigators, e.g. those working with pure cultures, and especially by those who study "steady state" operation of chemostat cultures. Samples taken and assayed for constituents give global mean values. The term "homogeneous culture" is often applied in these situations, but in such cultures there are at least three possible sources of heterogeneity (Fig. 1.1):

1. The first, arising from differences in individuals at different stages of their cell cycles, is extremely well documented (see Lloyd et al. (1982) for a list of references complete to 1980, and Cooper (1991) for the more recent literature on bacteria). Because cellular growth and division are to a large

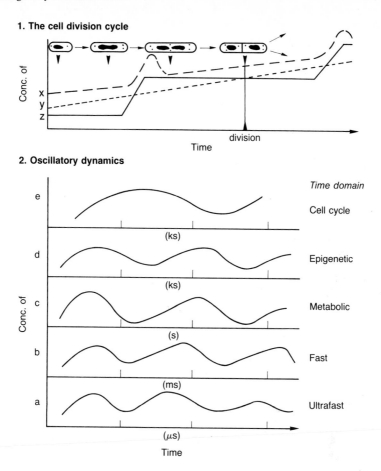

Fig. 1.1. Sources of heterogeneity in cultures.

extent discontinuous processes, data obtained from sampled populations are time-averaged over an interval equal to the division time. The details of sequences of processes occurring during the cell cycle are thereby hidden. Population heterogeneity may only be resolved by examination of each individual organism in turn, and flow cytometry provides an excellent facility for achieving this for a range of characteristics.

2. As well as the differences between organisms arising from their cell-cycle age differences, a second source of heterogeneity arises from oscillatory intracellular dynamics, e.g. ultradian clock-controlled properties (Lloyd and Stupfel 1991; Lloyd 1992). In lower eukaryotes many biochemical variables

are coupled to this timekeeper, which results in rhythmic operation (of energy metabolism, protein turnover, enzyme amounts and activities, etc.). In non-synchronized populations, individual organisms will be at different ultradian phases; values for the oscillatory periods (τ) observed lie within the range 30–90 min and are species-dependent. Biochemical oscillations with more rapid relaxation times (τ of the order of seconds) are also ubiquitous and thus provide a source of heterogeneity on shorter time scales. The most thoroughly investigated examples of metabolic oscillations are those of glycolysis (Das and Busse 1991), Ca^{2+} (Berridge 1989) and redox state (Visser et al. 1990), and those in *Dictyostelium discoideum* dependent on cyclic AMP (Goldbeter 1990).

3. Further types of heterogeneity within a genetically uniform population are expressed when organisms grow under unfavourable conditions. Sub-populations may then emerge. Various conditions of stress have been studied; both survival and recovery have marked effects on ultrastructure and biochemical function. Starvation may lead to the production of very small organisms (Novitsky and Morita 1976; Amy and Morita 1983). These tiny bacterial variants ($<0.2\,\mu m$ diameter) may easily escape attention or enumeration (Matin 1992). Pedigree analysis of cultured mammalian cell lines growing at low serum concentrations suggests that cell division times show wide dispersion only partly because of the randomness of acquisition of limiting nutrient molecules (Volkov et al. 1992). Environmental stress can induce bacteria to enter a resting stage (Bisset 1952) or to survive as "injured" organisms (Russell 1991). Such organisms have been variously termed "ultrabacterial cells" (Novitsky and Morita 1976), growth precursor cells (Dow et al. 1983), "viable but not culturable" (Rollins and Colwell 1986), "somnicells" (Rozak and Colwell 1987; Rozac et al. 1984), or "quiescent cells" (Lewis and Gattie 1991). Traditional plate count estimations, or even slide culture techniques (Powell 1956; Postgate et al. 1961), may not successfully reveal the presence of quiescent organisms, and ultra-microcolonies are easily overlooked. Quesnel (1960) commented on the heterogeneity of ability for colony formation within a bacterial population, and examples of the effects of diluent and media on colony counts have been documented (Mossel and Corry 1977). After a period of substrate deprivation, high concentrations of nutrients can prove lethal (substrate-accelerated death: Postgate and Hunter 1962, 1964).

Development of Flow Cytometers for Microbiology

The DNA content of a yeast is about 200-fold less than that of a typical mammalian cell, and the corresponding factor for *Escherichia coli* is 1400. It is therefore not surprising that the development of refined instruments capable of providing useful information on tiny organisms with small genomes (Table 1.1) is a comparatively recent event. Early efforts were devoted to the determination of frequency distribution curves for cellular components within randomly dividing populations of mammalian cells, and by 1965 flow cytometry began to replace autoradiography for the determination of the proportion of a cell population in S-phase (Kamentsky et al.

Table 1.1. Mean cell volumes and DNA content of some microorganisms

Organism	Typical cell volumes (μm^3) determined using Coulter counter	DNA content (fg)	Reference[a]
Tetrahymena pyriformis ST	8000[b]	210	Eisert et al. (1975)
Acanthamoeba castellanii Neff	1000[b]	3600	Coulson and Tyndall (1978)
Plasmodium yoelii		59	Jackson et al. (1977)
Trichomonas vaginalis	600[c]	1000	Alderete et al. (1986)
Schizosaccharomyces pombe 972h	60[b]	12	Agar and Bailey (1972)
Saccharomyces cerevisiae (haploid)	40[b]	25	Hutter and Eipel (1978)
Azotobacter vinelandii	2.7–7.4[d]		
Bacillus subtilis	1.3[e]		
Escherichia coli	0.45[f]	4.2	Steen and Boye (1980)

[a] For DNA contents; [b] Lloyd et al. (1977); [c] Paget and Lloyd (1990); [d] Allman et al. (1990); [e] Edwards and McCann (1981); [f] Scott et al. (1980).

1965). The simplicity and rapidity of the technique by comparison with traditional methods was astonishing, and the facility (e.g. using mithramycin in aqueous ethanol) for monitoring cell cycle kinetics during a continuing experiment (Crissman and Tobey 1974) encouraged its wide-spread use. The growing awareness of the technique in tissue-culture laboratories was not paralleled by microbiological applications, despite a very early demonstration of the use of light scatter to quantitate bacteria (Ferry et al. 1949). The publication of a paper (Arndt-Jovin and Jovin 1974) that showed that *E. coli* could be detected by light scattering in a flow system was a revelation. Reports on the flow cytometry of lower eukaryotes soon followed, such as those on *Saccharomyces cerevisiae* (Hutter et al. 1975a,b), *Tetrahymena pyriformis* (Eisert et al. 1975; Phillips and Lloyd 1978), and *Euglena gracilis* (Falchuk et al. 1975). Even viruses were shown to be detectable (Hercher et al. 1979). Attempts at about this time to count bacteria with an early version of the Ortho Flow Cytofluorimeter (E.W. Meyer and D. Lloyd, unpublished results 1975) indicated that the (then) only commercially available instrument in the UK was unsuitable, although extremely useful for cell cycle studies on yeasts and protozoa. Thus at that time flow cytometry could not challenge the Coulter counter for automated bacterial enumeration. Other instruments were soon to become available, and early investigations included the cell cycle kinetics of *S. cerevisiae* (Slater et al. 1977), the purity of yeast cultures (Hutter et al. 1978) and the "analysis" of bacterial and other microbial cultures (Paau et al. 1977; Hutter and Eipel 1978, 1979; Hutter and Oldiges 1980).

Resolved histograms of DNA and protein content of *Bacillus subtilis* were first obtained by Fazel-Madjlessi and Bailey (1979), but the real breakthrough in working with bacteria came with the invention by Steen and Lindmo (1979) of a completely new kind of instrument. This machine, the microscope-based flow cytometer, uses a hydrodynamically directed stream of cells passed across a coverslip through a focus of exciting light from a mercury arc lamp (Steen et al. 1989). Steen and Boye (1980, 1981) were able to measure the DNA content of individual bacteria (*E. coli*) with an accuracy of a few per cent, and they went on to study the DNA replication cycle at different growth rates (Steen et al. 1982; Skarstad et al. 1983, 1985). They were also able to obtain cells containing multiples of fully replicated chromosomes, using rifampicin which allows continuing rounds of replication to terminate but inhibits initiation (Skarstad et al. 1986). There followed from the Oslo group a series of seminal contributions on mutant bacteria with defective initiation proteins (Skarstad et al. 1988a,b; Boye et al. 1988, 1989).

Alongside these developments, modifications and optical improvements to the traditional laser-based instruments enabled other studies on, for example, bacterial plasmid heterogeneity (Dennis et al. 1983). Flow immunofluorescence techniques were used for the specific detection of *Bacillus anthracis* spores (Phillips et al. 1987), and for *Legionella pneumophila* and *E. coli* (Phillips and Martin 1988).

The scope of microbiological investigations using flow cytometry continues to expand and provide important new insights. The discovery of marine nanoplankton (Chisholm et al. 1988), now believed to be of major significance for marine primary productivity, and the ability to measure the viability of a pathogenic protozoan (*Giardia muris*) (Heyworth and Papo 1989) are just two of many recent achievements. The contributions to this volume highlight some important and exciting new vistas for flow cytometry in microbiology.

References

Agar DW, Bailey JE (1982) Cell cycle operation during batch growth of fission yeast populations. Cytometry 3:123

Alderete JF, Kasmala L, Metcalfe E, Garza GE (1986) Phenotypic variation and diversity among *Trichomonas vaginalis* isolates and correlation of phenotype with trichomonal virulence determinants. Infect Immunol 53:285

Allman R, Hann AC, Phillips AP, Martin KL, Lloyd D (1990) Growth of *Azotobacter vinelandii* with correlation of Counter cell size, flow cytometric parameters and ultrastructure. Cytometry 11:822–831

Amy PS, Morita RY (1983) Starvation survival patterns of 16 freshly isolated open ocean bacteria. Appl Environ Microbiol 45:1109–1115

Arndt-Jovin DJ, Jovin TM (1974) Computer-controlled cell (particle) analyser and separator: use of light scattering. FEBS Lett 44:247

Berridge MJ (1989) Cell signalling through cytoplasmic calcium oscillators. In: Goldbeter A (ed) Academic Press, London, pp. 449–460

Bisset KA (1952) Bacteria. Livingstone, Edinburgh

Boye E, Løbner-Olesen A (1990) Flow cytometry: illuminating microbiology. The New Biologist 2:119–125

Boye E, Løbner-Olesen A, Skarstad K (1988) Timing of chromosomal replication in *Escherichia coli*. Biochim Biophys Acta 951:359–364

Chisholm SW, Olson RJ, Zettler ER, Goericke R, Waterbury JB, Welschmeyer NA (1988) A novel free-living prochlorophyte abundant in the oceanic euphotic zone. Nature 334:340–343

Cooper S (1991) Bacterial growth and division. Academic Press, London

Coulson PB, Tyndall R (1978) Quantitation by flow microfluorimetry of total cellular DNA in *Acanthamoeba*. J Histochem Cytochem 26:713–718

Crissman H, Tobey RA (1974) Cell cycle analysis in 20 minutes. Science 184:1297–1298

Das J, Busse HG (1991) Analysis of the dynamics of relaxation type oscillation in glycolysis of yeast extracts. Biophys J 60:369–379

Dennis K, Srienc F, Bailey JE (1983) Flow cytometric analysis of plasmid heterogeneity in *Escherichia coli* populations. Biotechnol Bioeng 25:2485–2489

Dow CS, Whittenbury R, Carr NG (1983) The "shut down" or "growth precursor cell" – an adaptation for survival in a potentially hostile environment. Symp Soc Gen Microbiol 34:187–247, Cambridge University Press, Cambridge

Edwards C, McCann RJ (1981) Differential effects of inhibitors on respiratory activity of synchronous cultures of *Bacillus subtilis* prepared by continuous-flow centrifugation. J Gen Microbiol 125:47–53

Eisert WG, Ostertag R, Niemann EG (1975) Simple flow microphotometer for rapid cell population analysis. Rev Sci Instrum 46:1021

Falchuk KH, Krishan A, Vallee BL (1975) DNA distribution in the cell cycle of *Euglena gracilis*. Cytofluorometry of zinc deficient cells. Biochemistry 14:3439

Fazel-Madjlessi J, Bailey JE (1979) Analysis of fermentation process using flow microfluorimetry: single-parameter observations of batch bacterial growth. Biotech Bioeng 21:1995–2010

Ferry RM, Farr Jr LE, Hartman MG (1949) The preparation and measurement of the concentration of dilute bacterial aerosols. Chem Rev 44:389–395

Goldbeter A (1990) Rythmes et chaos dans les systèmes biochimique et cellulaires. Masson, Paris

Hayworth MF, Papo J (1989) Use of two-colour flow cytometry to assess killing of *Giardia muris* trophozoites by antibody and complement. Parasitology 99:199–203

Hercher M, Mueller W, Shapiro HM (1979) Detection and discrimination of individual viruses by flow cytometry. J Histochem Cytochem 27:350–352

Hill EC, Genner C (1981) Avoidance of microbial infection and corrosion in slow-speed diesel engines by improved design of the crankcase oil system. Tribology Int 8:67–74

Hutter KJ, Eipel HE (1978) Flow cytometric determinations of cellular substances in algae, bacteria, moulds and yeasts. Anton van Leeuw 44:269

Hutter KJ, Eipel HE (1979) Microbial determination by flow cytometry. J Gen Microbiol 113:369–375

Hutter KJ, Oldiges H (1980) Alterations of proliferating microorganisms by flow cytometric measurements after heavy metal intoxication. Ecotoxicol Environ Safety 4:57

Hutter KJ, Goehde W, Emeis CC (1975a) Investigation about the synthesis of DNA, RNA and proteins of selected populations of microorganisms by cytophotometry and pulse-cytophotometry. I. Methodical investigations about appropriate fluorescence-dyes and staining procedures. Chem Mikrobiol Technol Lebensum 4:29–32

Hutter KJ, Otto F, Emeis CC (1975b) Investigations about the synthesis of DNA, RNA and proteins of selected populations of microorganisms by cytophotometry and pulse-cytophotometry. II. Synthesis of DNA, RNA and proteins of yeast of the species *Saccharomyces* during the vegetative growth. Chem Mikrobiol Technol Lebensum 4:75–80

Hutter KJ, Eipel HE, Hettwer H (1978) Rapid determination of the purity of yeast cultures by flow cytometry. European J Appl Microbiol Biotechnol 5:109–112

Jackson PR, Winkler DG, Kimzey SL, Fisher FM Jr (1977) Cytofluorograf detection of *Plasmodium yoelii*, *Trypanosoma gambiense*, and *Trypanosoma equiperdum* by laser excited fluorescence and stained rodent blood. J Parasitol 63:593–598

Kamentsky LA, Melamed MR, Derman H (1965) Spectrophotometer: new instrument for ultrarapid cell analysis. Science 150:630–631

Lewis DL, Gattie DK (1991) The ecology of quiescent microbes. Am Soc Microbiol 57:27–32

Lloyd D (1992) Intracellular timekeeping: epigenetic oscillations reveal functions of an ultradian clock. In: Lloyd D, Rossi ER (eds) Ultradian rhythms in life processess: A fundamental inquiry into chronobiology, Springer, London, pp 5–21

Lloyd D, Stupfel M (1991) The occurrence and function of ultradian rhythms. Biol Rev 66:275–299

Lloyd D, John L, Hamill M, Phillips C, Kader J, Edwards SW (1977) Continuous flow cell cycle fractionation of eukaryotic microorganisms. J Gen Microbiol 99:223–227

Lloyd D, Poole RK, Edwards SW (1982) The cell division cycle: temporal organization and control of cellular growth and reproduction. Academic Press, London

Matin A (1992) Physiology, molecular biology and applications of the bacterial starvation response. J Appl Bacteriol Suppl 73:49S–57S

Mossel DAA, Corry JEL (1977) Detection and enumeration of sublethally injured pathogenic and index bacteria in foods and water processed for safety. Kult und Diff 19–34

Novitsky JA, Morita RY (1976) Morphological characterization of small cells resulting from nutrient starvation of a psychotrophic marine vibrio. Appl Environ Microbiol 32:616–622

Paau AS, Cowles JR, Oro J (1977) Flow-microfluorimetric analysis of *Escherichia coli*, *Rhizobium melilotti*, and *Rhizobium japonicum* at different stages of the growth cycle. Can J Microbiol 23:1165–1169

Paget TA, Lloyd D (1990) *Trichomonas vaginalis* requires traces of oxygen and high concentrations of carbon dioxide for optimal growth. Mol Biochem Parasitol 41:65–72

Peters DC (1979) A comparison of mercury arc lamp and laser illumination for flow cytometry. J Histochem Cytochem 27:241–245

Phillips AP, Martin KL (1988) Limitations of flow cytometry for the specific detection of bacteria in mixed populations. J Immunol Meth 106:109–117

Phillips AP, Martin KL, Capey AJ (1987) Direct and indirect immunofluorescence analysis of bacterial populations by flow cytometry. J Immunol Meth 101:219–228

Phillips CA, Lloyd D (1978) Continuous-flow size selection of *Tetrahymena pyriformis* ST: changes in volume, DNA, RNA and protein during synchronous growth. J Gen Microbiol 105:95–103

Postgate JR, Hunter JR (1962) The survival of starved bacteria. J Gen Microbiol 29:233–263

Postgate JR, Hunter JR (1964) Accelerated death of *Aerobacter aerogenes* starved in the presence of growth limiting substrates. J Gen Microbiol 34:459–473

Postgate JR, Crumpton JE, Hunter JR (1961) The measurement of bacterial viabilities by slide culture. J Gen Microbiol 24:15–24

Powell EO (1956) A rapid method for determining the proportion of viable bacteria in a culture. J Gen Microbiol 14:153–159

Quesnel (1960) The behaviour of individual organisms in the lag phase and the development of small populations of *Escherichia coli*. J Appl Bacteriol 23:99–105

Rollins DM, Colwell RR (1986) Viable but non-culturable stage of *Campylobacter jejuni* and its role in survival in the natural aquatic environment. Appl Environ Microbiol 52:531–538

Roszak DB, Colwell RR (1987) Metabolic activity of bacterial cells enumerated by direct viable count. Appl Environ Microbiol 53:2889–2983

Roszak DB, Grimes DJ, Colwell RR (1984) Viable but non-recoverable stage of *Salmonella enteritidis* in aquatic systems. Can J Microbiol 30:334–338

Russell AD (1991) Injured bacteria: occurrence and possible significance. Lett Appl Microbiol 12:1–2

Scott RI, Gibson JF, Poole RK (1980) Adenosine triphosphatase activity and its sensitivity to Ruthenium Red oscillate during the cell cycle of *Escherichia coli* K12. J Gen Microbiol 120:183–198

Shapiro HM (1983) Multistation multiparameter flow cytometry: a critical review and rationale. Cytometry 3:227–243

Shapiro HM (1985) The little laser that could: applications of low power lasers in clinical flow cytometry. Ann New York Acad Sci 468:18–27

Skarstad K, Boye E (1988) Perturbed chromosomal replication in recA mutants of *Escherichia coli*. J Bacteriol 170:2549–2554

Skarstad K, Steen HB, Boye E (1983) Cell cycle parameters of slowly, growing *E. coli* B/r studied by flow cytometry. J Bacteriol 154:656–662

Skarstad K, Steen HB, Boye E (1985) DNA distributions of *E. coli* measured by flow cytometry and compared to theoretical computer simulations. J Bacteriol 163:661–668

Skarstad K, Boye E, Steen HB (1986) Timing of initiation of chromosome replication in individual *Escherichia coli* cells. EMBO J 5:1711–1717

Skarstad K, von Meyenburg K, Hansen FG, Boye E (1988) Coordination of chromosome replication initiation in *Escherichia coli*: effects of different *dnaA* alleles. J Bacteriol 170:852–858

Skarstad K, Løbner-Olesen A, Atlung T, von Meyenburg K, Boye E (1989) Initiation of DNA replication in *Escherichia coli* after overproduction of the DnaA protein. Mol Gen Genet 218:50–56

Slater ML, Sharrow SO, Gart JJ (1977) Cell cycle of *Saccharomyces cerevisiae* in populations growing at different rates. Proc Natl Acad Sci USA 74:3850–3854

Steen HB, Boye E (1980) Bacterial growth studied by flow cytometry. Cytometry 1:32–36

Steen HB, Boye E (1981) Growth of *Escherichia coli* studied by dual parameter flow cytometry. J Bacteriol 145:1091–1094

Steen HB, Lindmo T (1979) Flow cytometry: a high resolution instrument for everyone. Science 204:403–404

Steen HB, Boye E, Skarstad K, Bloom B, Godal T, Mustafa S (1982) Applications of flow cytometry on bacteria: cell cycle kinetics, drug effects and quantitation of antibody binding. Cytometry 2:249–257

Steen HB, Lindmo T, Stokke T (1989) Differential light-scattering detection in an arc lamp-based flow cytometer. In: Yen A (ed) Flow cytometry: Advanced research and clinical applications, vol. I. CRC Press, Boca Raton, Florida, p 63

Visser G, Reinten C, Coplan P, Gilbert DA, Hammond K (1990) Oscillations in cell morphology and redox state. Biophys Chem 37:383–394

Volkov EI, Stolyarov MN, Brooks RF (1992) The modelling of heterogeneity in proliferative capacity during clonal growth. In: Volkov E (ed) Biophysical approach to complex biological phenomena. Nova, New York, pp 183–203

Chapter 2

The Physical and Biological Basis for Flow Cytometry of *Escherichia coli*

Erik Boye and Harald B. Steen

Introduction

Flow cytometry is a method for measuring the fluorescence and light scattering of individual cells in large numbers. By labelling the cells with fluorescent molecules that bind with high specificity to one particular cellular constituent it is possible to measure the content of the constituent. Such a fluorescent tag may be either a dye molecule with a high binding specificity for the particular component to be measured or a fluorescence conjugated antibody. The light scattering of the cells gives information on their size and to some extent also on shape and structure (Salzman et al. 1979, 1990).

In the flow cytometer the cells are carried by a laminar flow of water through a focus of light, the wavelength of which matches as closely as possible the absorption spectrum of the dye with which the cells have been stained. On passing through the focus, each cell emits a pulse of fluorescence and scattered light that is collected by means of lenses and directed onto sensitive detectors. These detectors, which are usually photomultiplier tubes (PMT), transform the light pulses into equivalent electrical pulses, which are measured and digitized by appropriate electronics before the data are stored in a computer. The fluorescence may be split into different colour components, so that several different dyes can be measured simultaneously by separate detectors. Light scattered to small and large angles may be measured by separate detectors so that cells can be distinguished on the basis of size as well as structure.

With such an instrument, it is possible to measure various cell parameters for several thousand cells per second with a precision of a few per cent or less and with a sensitivity allowing detection of a few hundred fluorescent molecules per cell. Not surprisingly, this technique has found numerous applications in various fields of cell biology and medicine.

In spite of the fact that flow cytometers have been commercially available for about two decades they have been used almost exclusively for measurements of mammalian, or at least eukaryotic, cells, while they have remained a rarity in microbiology, including bacteriology. The most obvious reason for this is that the volume of a bacterium is roughly three orders of magnitude smaller than that of a mammalian cell, and so are the amounts of various cellular constituents. For example, bacterial DNA has a mass in the order of 3×10^{-15} g per cell and the amount of dye that binds to it may be one order of magnitude smaller. Thus, flow cytometry of bacteria is highly demanding with regard to the sensitivity and measuring precision of the instrument (Steen 1990a). In the following we shall discuss some of the factors that determine the performance of flow cytometers before we go on to give some practical examples of applications regarding control of cell growth and of DNA replication. For reviews of flow cytometric techniques in microbiology the reader is also referred to Steen (1990a), Steen et al. (1990) and Boye and Løbner-Olesen (1990).

The Physical Basis of Flow Cytometry

In order to generate a measurable fluorescence and light scattering signal from bacteria the instrument has to be optimized with regard to sensitivity. In general, sensitivity increases with the intensity of the exciting light. To achieve a high excitation intensity, light from a powerful light source, such as a laser or a high-pressure arc lamp, is concentrated into a very small focus. However, there is a limit to how small this focus can be made without sacrificing measuring precision and reliability. Thus, in order to obtain reproducible measurements all cells have to be exposed to the same excitation intensity, which means that the width of the path that the cells follow through the focus must be much smaller than the width of the focus. To make the cells follow the same path as reproducibly as possible one is utilizing "hydrodynamic focussing" of the sample, the principle of which is shown in Fig. 2.1. Water under pressure flows in a laminar fashion through a conical nozzle with an exit orifice of diameter in the order of 100 μm, which may lead either into open air, in which case the water forms a laminar jet of circular cross section, or into a quartz tube that typically has a rectangular cross section of somewhat larger dimensions (e.g. 250 μm square). This main flow of water, which acts as a carrier of the sample, is called the "sheath flow". The sample is introduced through a tube running along the centre of the nozzle. The diameter, d, of the sample flow is given by:

$$d = (s/S)^{\frac{1}{2}} \cdot D \qquad (2.1)$$

where s and S are the flow rate of the sample and the sheath, respectively, and D the diameter of the nozzle (Steen 1990b). Since the flow is laminar and the sample and the sheath move with a common velocity, the cross section of the sample flow relative to that of the nozzle will be maintained as the water runs down the conical nozzle and into the orifice. Thus, the sample will be confined to the central core of the flow through the focus of excitation light. To obtain reproducible measurements, the width of the focus has to be so large that the intensity is fairly constant across the width of the sample flow. Assuming that the focus has a Gaussian intensity profile

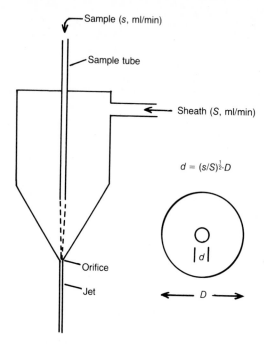

Fig. 2.1. The principle of hydrodynamic focussing. Water under pressure fills a conical nozzle and flows in a laminar way through the orifice forming a laminar jet. The sample is introduced through the tube along the central axis of the nozzle. Since the sample acquires the same velocity as the water, the diameter of the sample flow relative to that of the water will remain constant from the outlet of the sample tube and into the jet. As a result the sample is confined to a narrow central core of the jet.

this means that its width (between the points of $1/e^2$ of maximum intensity) must be about 15 times larger than that of the sample flow if the excitation intensity is to be constant to within 1%.

According to Eq. (2.1), a larger sample flow rate implies a larger diameter of the sample flow and thereby requires a wider focus if the measuring precision is to be maintained. Hence, a compromise must be found between the requirement for high excitation intensity on the one hand and large sample flow on the other. The typical result of such a trade-off is a focal width of about $100\,\mu$m, a sheath flow rate of about 10 ml/min and a sample flow rate that is limited to a maximum of $50\,\mu$l/min.

The velocity of the flow through the focus is in the order of 5 m/s, which means that the cell traverses the focus in $10\,\mu$s or less. This period of traverse limits the rate of measurement: since the cells arrive randomly distributed in time there is a certain probability that two cells will be in the focus within this time period, t, and therefore be detected as one cell. This probability or rate of coincidence, P, is given by:

$$P = t \cdot r \tag{2.2}$$

where r is the rate of measurement. For example, if $t = 10\,\mu$s and P is to be limited to 1%, then $r = 10^3$ cells/s.

The measuring precision, i.e. the reproducibility with which identical cells can be measured, is limited by the magnitude of the light signal, i.e. the number of photons reaching the light detector from each cell. The reason for this is to be found in a basic fact of nature, namely that emission of light is a stochastic process, which means that photons are emitted randomly in time. A consequence of this is that if we measure a series of identical cells in a flow cytometer, which in itself introduces no variability in the measurement, the average number of photons, n, in the signal reaching the detector during the measurement of each cell will vary with a standard variation, s, given by:

$$s = n^{\frac{1}{2}} \tag{2.3}$$

And thus by a relative coefficient of variation:

$$cv = n^{-\frac{1}{2}} \tag{2.4}$$

Hence, measuring precision is limited by the magnitude of the signal, although in some situations the signal may be so large that other sources of variation are dominant.

The light emitted by the cell itself is superimposed on a constant background of light, which is caused by light scattering and fluorescence from the flow chamber and the various lenses and other components in the optical path. As shown elsewhere (Steen 1991), the magnitude of this background is important for the sensitivity as well as the measuring precision of flow cytometers. Assuming that we measure a sample of identical cells, each containing f fluorescent molecules of a given type, the relative standard variation of the measurement of this fluorescence is given by:

$$cv = c[(f + b)/f^2 \cdot i_x]^{\frac{1}{2}} \tag{2.5}$$

where c is a numerical constant and b is an equivalent number of fluorescent molecules that would give rise to the actual level of background (Steen 1992). The cv is a measure of the width of the narrowest histogram peak that can be obtained for a signal of a given magnitude. It can be seen that the measuring precision is limited by the cellular content of fluorescent molecules as well as by the background. The lowest detectable number of a given type of fluorescent molecule, f_1, is (Steen, unpublished data):

$$f_1 = c(b/i_x)^{\frac{1}{2}} \tag{2.6}$$

Thus, it can be seen that the fluorescence sensitivity of flow cytometers is limited as much by the background as by the intensity of the excitation light. In instruments employing a closed flow chamber the background is partly due to dust and other impurities sticking to its surfaces.

Both cv and f_1 are inversely proportional to the square root of the excitation intensity. Hence, replacing the excitation light source by one with 10 times more power will increase sensitivity and measuring precision by a factor of about 3. In some cases a similar improvement in performance may be obtained simply by cleaning the flow chamber.

It should be noted that Eqs. (2.5) and (2.6) are based on the assumption that the intensity of the fluorescence as well as background is proportional to that of the excitation light. This is not necessarily the case, especially with regard to the fluorescence. When the excitation intensity exceeds about 100 mW "bleaching", i.e. photochemical degradation of the dye and/or saturation of its excited state, may significantly reduce the fluorescence yield (Pinkel et al. 1979; van den Engh and Farmer 1992). These phenomena are strongly dependent on the size of the focus, but are important only in the large laser-based instruments.

The constant in Eqs. (2.5) and (2.6) is given by:

$$c = \text{const} \cdot (\text{NA})^2 \cdot T \cdot v^{-1} \cdot \alpha \cdot \int \varepsilon(\lambda) \cdot I_x(\lambda) d\lambda \qquad (2.7)$$

where NA is the numerical aperture of the optics which collects the fluorescence, T the transmission of the fluorescence optical path, v the flow velocity, α the sensitivity of the light detector, and $\int \varepsilon(\lambda) \cdot I_x(\lambda) d\lambda$ the overlap between the absorption spectrum of the fluorescent dye and the spectrum of the excitation light.

The integral in Eq. (2.7) implies that sensitivity and measuring precision depend on the match between the absorption spectrum of the fluorescent dye and the wavelength of excitation. It also follows that these performance parameters can be improved as much by reducing the flow velocity, v, as by increasing the intensity of the excitation light source, which is usually much more complicated and expensive. It should be noted, however, that since with a given orifice of the flow chamber the sheath flow, S, is proportional to v, a reduction of v implies an increase in the sample flow diameter, d, according to Eq. (2.1), and thereby a possible reduction in measuring precision.

The Instrument

Flow cytometers can be divided into two major groups, depending on the type of excitation light source: instruments using an argon laser and those employing a high-pressure mercury or xenon lamp. An instrument of the latter type is shown in Fig. 2.2. The optical outline of the instrument resembles that of fluorescence microscopes with epi-illumination. The excitation light source is a 100 W mercury lamp, the arc of which is imaged on the "excitation slit", which in turn is imaged in the object plane of a microscope lens. The white light from the arc lamp is passed through the "excitation filter", which is an interference filter transmitting the wavelength used to excite the dye in question. This light is reflected via a dichroic mirror into a microscope lens that concentrates the light in the focus through which the sample flow is passing. Fluorescence from the cells is collected by the same lens and passes through the dichroic mirror (because of its longer wavelength) and onto the "emission slit", which is situated in the image plane of the microscope lens. Hence, the fluorescence is collected in the direction opposite to that of the excitation. Since this is the direction where the intensity of the stray light from the flow chamber as well as from the lens itself is the smallest, this optical configuration produces the lowest possible level of background. In order to concentrate a maximum intensity of excitation light in the focus and to collect as much of the cell fluorescence as

a

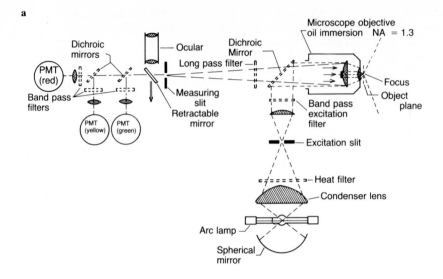

b

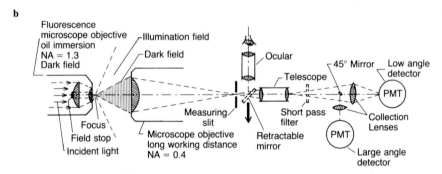

Fig. 2.2a,b. Arc-lamp-based flow cytometer. **a** Fluorescence is detected in epi-mode, which means that the same microscope objective is employed to concentrate the excitation light on the sample flow, which runs through its focus, and to collect fluorescence. This microscope objective forms a fluorescence image of the sample flow on the emission slit, which thereby eliminates background light from other parts of the image. The excitation and fluorescence wavelengths are determined by the combination of excitation filter, dichroic mirror and long pass filter, which may be chosen to give optimal fluorescence output from virtually any dye used in flow cytometry. **b** Light scattering is detected in a dark field configuration. A secondary microscope collects scattered light within the dark field produced by a field stop in the primary microscope objective. The dark field is reproduced behind the measuring slit by means of a telescope in order to facilitate separate detection of light scattered to small and large angles.

possible, the lens is a microscope objective with oil immersion and a numerical aperture close to the theoretical maximum, i.e. NA = 1.3.

The purpose of the excitation and emission slits is to eliminate stray light and thereby reduce the amount of background that is reaching the detectors. The excitation slit limits the region of the object plane that is illuminated so as to include only the sample flow and its immediate surroundings, whereas the emission slit, which is set to cover essentially the same region, eliminates light emitted by sources other than the sample flow. The efficiency of this so-called spatial filtering depends on the resolution and contrast of the image of the excitation slit in the focal plane and of the fluorescence image of the sample flow on the emission slit. The use of high-quality microscope optics is important in this regard.

Light scattering detection is carried out in a dark field configuration (Steen et al. 1989). The microscope lens contains a central field stop that produces a conical shadow pointed in the focus and extending on the opposite side. This shadow contains only light originating in the focus, that is, fluorescence and scattered light. Hence, a lens having its object plane in common with the first lens and its aperture within this shadow, "sees" only fluorescence and scattered light. The fluorescence, which is generally much weaker than the scattered light, can be eliminated by an appropriate filter, so that the secondary lens is forming a scattered light image of the cell flow on an emission slit, which eliminates stray light and reduces background by the same principle as in the fluorescence detection. In the scattered light field behind the slit it is possible to isolate light scattered to low and high angles, respectively, so that these scattering components can be measured by separate detectors (Steen 1990c).

The flow chamber is of the "jet on open surface" type. The jet from a nozzle with hydrodynamic focussing impinges on the open surface of a cover slip so as to produce a flat, laminar flow on this surface with the cells confined to a narrow sector along the middle of the flow. This type of flow chamber exhibits fewer interfaces than closed flow chambers, and therefore gives rise to less stray light and a correspondingly lower background (Steen 1990b).

The sample injection system is an important part of the flow cytometer. In most commercial instruments this system is of the "differential pressure" type, which implies that the sample flow rate is not exactly known. We prefer the "volumetric injection" type where the sample is injected into the nozzle from a syringe the plunger of which is moved at a calibrated rate. Thus, the sample flow is known and the cell concentration can be determined directly from the number of cells detected per unit time (Steen 1990b).

The near-parallel excitation light used in laser-based flow cytometers may create artifacts in the measurement of non-spherical cells such as erythrocytes or rod-like bacteria, since the fluorescence as well as the light scattering will depend on the orientation of the cell relative to the direction of the laser beam. In arc-lamp-based instruments with microscope optics the cone of excitation light is very much wider and the artifact therefore reduced significantly.

DNA Replication in Bacteria

Most bacteria contain a single, circular chromosome that is replicated bidirectionally from a unique origin of replication. By varying nutrient availability in the medium, bacteria may be grown at a wide range of growth rates, accompanied by large variations in the cell size and DNA content.

DNA replication is coupled to general cell growth by some as yet unidentifed mechanism(s). It is a most important undertaking to unravel these regulatory mechanisms, and their revelation will have far-reaching consequences. In this work, flow cytometry will be of prime importance.

It takes an *Escherichia coli* cell 40–55 min (the C period) to replicate the chromosome at rapid growth rates, somewhat depending on the strain (Cooper and Helmstetter 1968; Skarstad et al. 1985; Allman et al. 1991). After termination of DNA replication, cell division ensues within a period of about 25 min (the D period). In rich growth medium cells of *E. coli* may divide every 20 min, which means that the doubling time is less than one-third of the duration of the DNA replication cycle. To be able to continue dividing every 20 min the bacteria must initiate a new round of DNA replication every 20 min, i.e. before the previous round of replication is finished. Thus, the chromosome of rapidly growing *E. coli* cells is a multi-forked, complex structure. Termination of one round of replication on a multiforked chromosome results in a separation into two equal parts that are segregated and partitioned at division, one into each daughter cell.

Initiation of DNA Replication

DNA replication in *E. coli* is initiated at the unique origin, termed *oriC*, and proceeds bidirectionally (Masters and Broda 1971; Bird et al. 1972) around the chromosome until the forks reach the terminus region, *terC*, diametrically opposite *oriC*. Genetic experiments have identified the DnaA protein as specifically required for the initiation step, and biochemical data show that DnaA binds *oriC* and separates the strands of the DNA helix to allow replication proteins to start semiconservative DNA synthesis (for a review, see von Meyenburg and Hansen 1987). Before the action of DnaA in initiation both protein synthesis *de novo* (Pritchard and Zaritsky 1970) and a transcriptional event (Lark 1972; Messer 1972) are required. Thus, inhibitors of protein synthesis and of transcription (RNA polymerase) inhibit initiation of DNA replication, but the nature of the required protein or transcriptional event is not known. In the presence of chloramphenicol (protein synthesis inhibitor) or rifampicin (RNA polymerase inhibitor) initiations are blocked, but ongoing rounds of replication are not impeded in their progression towards the termini.

Regulation of DNA Replication

The time it takes a replication fork to proceed from *oriC* to *terC* (the C period) has been reported to be independent of cell growth rate, except at very slow growth (Cooper and Helmstetter 1968), and the rate of DNA replication has therefore been assumed to be regulated at the level of initiation (von Meyenburg and Hansen 1987). More recently, in experiments involving a sudden increase in the intracellular DnaA concentration, repli-cation forks were shown to be slowed down and even stopped permanently at some distance from *oriC* (Atlung et al. 1987; Skarstad et al. 1989; Løbner-Olesen et al. 1989). The concept of a constant C period has not been challenged with critical experimentation, and the possibility must be kept open that the C period duration actually varies with growth rate. Certainly, the length of the C period may vary significantly from strain to strain (Allman et al. 1991).

By combining the time of initiation in *E. coli* cells at different growth rates (Cooper and Helmstetter 1968) with cell size measurements of the related bacterium *Salmonella typhimurium*, it was deduced that the initiation mass was constant and independent of the growth rate (Donachie 1968). Initiation mass was defined as the cell mass at initiation divided by the number of origins to be initiated. Thus, it seemed that the cells needed to accumulate a certain mass per origin to be able to initiate replication. The concept of a constant initiation mass has served a valuable function in the study of initiation control for a number of years. However, exact measurements of the initiation mass at different growth rates have suggested that it may actually vary with growth rate (Churchward et al. 1981; Wold et al. unpublished data), and the initiation mass may therefore not be an adequate concept to use in the search for regulatory mechanisms.

When the intracellular concentration of DnaA is increased, as achieved by induction of an inducible *dnaA* gene on a plasmid, extra initiations can be observed (Atlung et al. 1987; Pierucci et al. 1987; Xu and Bremer 1988; Skarstad et al. 1989). This demonstrates that the DnaA protein is normally limiting initiation. Furthermore, in the presence of increased or decreased steady-state concentrations of DnaA protein initiation occurs earlier or later in the cell cycle, respectively (Løbner-Olesen et al. 1989). The time of initiation is determined by the DnaA concentration, but only within certain limits (Skarstad et al. 1989; Løbner-Olesen et al. 1989).

DNA Contents

The DNA contents of individual bacteria can be determined by the use of DNA-specific chromophores such as propidium iodide, Hoechst 33258, DAPI (4,6-diamino-2-phenylindole) or mithramycin (Shapiro 1988). As described below, the majority of our flow cytometric studies of bacteria have been concentrated on measurements of DNA. Since bacteria under normal growth conditions contain relatively much more RNA than mammalian cells, it is essential in such measurements to use dyes which combine good specificity for DNA with high fluorescence efficiency. Hence, dyes which bind to both DNA and RNA, such as ethidium bromide and propidium iodide, are not suitable alone. We have found that a combination of mithramycin and ethidium bromide gives the best results (Steen et al. 1990). In this combination mithramycin is the DNA-specific component and the excitation wavelength must be that of mithramycin, i.e. around 440 nm. Whereas the argon laser has no major emission line to match the absorption spectrum of mithramycin the mercury lamp exhibits a strong emission line at 436 nm. Ethidium bromide, which does not absorb at this wavelength, serves to enhance the yield of fluorescence by receiving excitation energy from

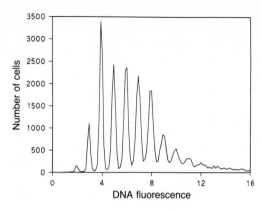

Fig. 2.3. DNA histogram of bacteria containing fully replicated chromosomes. *Escherichia coli* strain CM742 *dnaA*46 was grown in rich medium at 30 °C and treated with rifampicin for 3 h before fixation in ethanol, and staining with mithramycin and ethidium bromide before flow cytometry. This *dnaA* mutant cannot perform synchronous initiations, and the number of origins is therefore different from 2^n. (Reproduced from Boye and Løbner-Olesen (1991), with permission.)

adjacent mithramycin molecules. The DNA contents of bacteria may be accurately quantitated by the use of this drug combination, as exemplified by the low cv of 2.5% obtained for *E. coli* cells containing fully replicated chromosomes (Fig. 2.3).

Replication Origins

As noted above, *E. coli* cells incubated in the presence of rifampicin (RIF) complete all ongoing rounds of replication and end up with fully replicated, circular chromosomes (Fig. 2.3). Each chromosome contains one origin, which had to be present also at the time of drug addition (since RIF inhibits initiation). Thus, by counting the number of fully replicated chromosomes per cell the number of origins may be measured. However, this is not equal to the number of origins present at the time of drug addition unless cell division is stopped. RIF inhibits cell division, but only after a time delay. To measure the number of origins and eliminate the uncertainty of cell division, the rapidly acting division inhibitor cephalexin (CPX) may be added together with RIF (Boye and Løbner-Olesen 1991).

Synchrony of Initiation

After treatment with RIF, most *E. coli* strains contain 2^n fully replicated chromosomes, where $n = 0, 1, 2, 3$, etc. It may be concluded that an initiation event doubles the number of origins present per cell, meaning that all origins are initiated simultaneously (Skarstad et al. 1986). Initiation

synchrony depends on the action of several different gene products. Certain
dnaA(Ts) mutants lose the ability to perform synchronous initiations, even
at permissive temperatures (Skarstad et al. 1988). This conclusion can be
drawn from histograms of RIF-treated mutants, which may contain large
populations of cells with three, five, six or seven origins (see example in Fig.
2.3). The molecular nature of the defect is not known, but the mutations
conferring the asynchronous phenotype all affect the putative ATP-binding
site of the DnaA protein.

Other gene products that are necessary for synchronous initiations are
RecA (Skarstad and Boye 1988), RpoC (Boye et al. 1988) and Dam (Boye
et al. 1988; Boye and Løbner-Olesen 1990). The asynchrony discovered
for initiation in cells deficient in Dam methyltransferase may reflect a
hemimethylation-dependent binding of *oriC* after initiation, to prevent
multiple initiations in the same cell cycle (Boye 1991).

Cell Age at Initiation

If we can measure the number of origins of all the cells in an exponentially
growing population, cell age at the time of initiation may be determined. A
population of cells with synchronous initiations, treated with RIF/CPX, is
divided into two separate populations in a DNA histogram (Fig. 2.4a). The
left-hand peak represents cells that have not yet initiated replication, while
the right-hand peak represents cells with twice as many chromosomes (and
origins), where initiation of replication has occurred in that cell cycle. The
time of initiation (t_i) can be found from the age distribution (Fig. 2.4b),
where the fraction of cells before initiation is the same as in Fig. 2.4a
(hatched). With the same logic, cell mass at initiation (M_{t_i}) can be found
from the cell mass (light scatter, see below) distribution of cells not treated
with RIF/CPX (Fig. 2.4c).

This method of determining the time of initiation and initiation mass is
based on two provisions: First, the drugs must act immediately, or at least
simultaneously. CPX action is indeed very rapid (Boye and Løbner-Olesen
1991), but it is not clear how quickly RIF inhibits initiation. Second, for the
initiation mass measurements, the relevant time point is when the forks
actually leave the *oriC* region, which might be different from the time when
initiation passes the RIF-sensitive step. Thus, the measurements must be
interpreted with caution until the kinetics of drug action in different strains
and at different growth rates have been established.

Cell Mass

In flow cytometry the classical measure of cell mass is scattered light.
Scattering by small particles, such as bacteria, is a complex function of size,
shape and refractive indices (Salzman et al. 1979, 1990). The amount of light
scattered in the forward direction (less than 15°) is primarily determined by
the size of the cell, whereas at larger angles the intensity depends to a

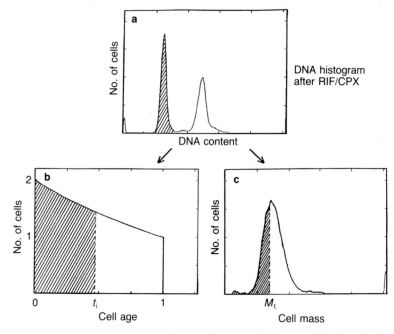

Fig. 2.4a–c. Method for determining cell age and mass at initiation of DNA replication. A DNA histogram of cells in a culture treated with RIF and CPX is shown in **a**. The cells with the lowest amount of DNA (hatched peak to the left) have not yet initiated replication. These cells can also be assumed to be both the youngest and the smallest in the population. The hatched portion in **a** therefore represents the same cells as the hatched fraction in the age distribution in **b**. The cells initiate replication at the age t_i. Also, the same fraction can be found among the smallest cells in the cell size distribution (**c**), where M_{t_i} is the mass at initiation.

greater extent on the shape and structure of the cell as well as its refractive index.

Cell Protein

Cellular contents of protein may be measured after staining with fluorescein isothiocyanate (FITC), which is a fluorescent dye giving quantitative staining of protein (Freeman and Crissman 1975). Within one population of exponentially growing cells, the correlation between low-angle light scattering and FITC fluorescence is strictly linear (B. Voss et al., unpublished data). The relationship between the two parameters over a wider range of variation in cell size is not yet known.

General Cell Growth

With its possibilities for exact measurements of cellular DNA content and of cell size or protein content, flow cytometry is an ideal technique for monitoring bacterial growth. Just as initiation of replication causes a dramatic (twofold) increase in DNA content after RIF/CPX treatment, so cell division causes a twofold reduction in cell mass. Thus, flow cytometry holds great promise for the study of cell division as a result of its demonstrated potential in DNA replication studies.

Exponential growth of a bacterial culture, as measured by an increase in the optical density (OD), continues to an OD of between 0.5 and 1.0 before growth rate decreases. Detailed flow cytometric investigations have shown that steady-state growth, i.e. a state where on average all cellular parameters in the culture remain the same, stops before deviations can be seen from the OD measurements (Skarstad et al. 1983; Steen 1990a). Flow cytometry appears to be a very sensitive method for monitoring cell growth, since the shape of the DNA histogram is extremely sensitive to changes in the growth conditions.

Counting Bacteria by Flow Cytometry

The classical way of determining the number of bacteria in a given volume is to dilute the bacteria on a solid agar surface and score colony formation or to count the number of particles in a Coulter counter. The former method gives a result only after 1 day of incubation and is hampered by the fact that bacteria do not always form colonies, and also by the tedious procedure involved if good counting statistics are required. Coulter counting is fairly simple in principle, but the necessity of using small orifices to measure bacteria often introduces several technical problems. Since a flow cytometer detects bacteria as fluorescence or light scatter signals it may conveniently be used as an exact and sensitive counter of bacteria (Steen 1990b; Boye and Løbner-Olesen 1991). To use the flow cytometer as a counter it is necessary that the sample flow rate is known, as described above.

References

Allman R, Schjerven T, Boye E (1991) Cell cycle parameters of *Escherichia coli* K-12. J Bacteriol 173:7970–7974

Atlung T, Løbner-Olesen A, Hansen FG (1987) Overproduction of DnaA protein stimulates initiation of chromosome and minichromosome replication in *Escherichia coli*. Mol Gen Genet 296:51–59

Bird R, Louarn J, Martuscelli J, Caro L (1972) Origin and sequence of chromosome replication in *Escherichia coli*. J Mol Biol 70:549–566

Boye E (1991) A turnstile for initiation of DNA replication. Trends Cell Biol 1:107–109

Boye E, Løbner-Olesen A (1990) Flow cytometry: illuminating microbiology. New Biol 2:119–125

Boye E, Løbner-Olesen A (1991) Bacterial growth control studied by flow cytometry. Res Microbiol 142:131–135

Boye E, Løbner-Olesen A, Skarstad K (1988) Timing of chromosomal replication in *Escherichia coli*. Biochim Biophys Acta 951:359–364

Churchward G, Estiva E, Bremer H (1981) Growth rate-dependent control of chromosome replication initiation in *Escherichia coli*. J Bacteriol 145:1232–1238

Cooper S, Helmstetter CE (1968) Chromosome replication and the division cycle of *Escherichia coli* B/r. J Mol Biol 31:519–540

Donachie WD (1968) Relationship between cell size and time of initiation of DNA replication. Nature 219:1977–1079

Freeman DA, Crissman HA (1975) Evaluation of six fluorescent protein stains for use in flow microfluorometry. Stains Technol 50:279–284

Lark KG (1972) Evidence for direct involvement of RNA in the initiation of DNA replication in *E. coli* 15T⁻. J Mol Biol 64:47–60

Løbner-Olesen A, Skarstad K, Hansen FG, von Meyenburg K, Boye E (1989) The DnaA protein determines the initiation mass of *Escherichia coli* K-12. Cell 57:881–889

Masters M, Broda P (1971) Evidence for the bidirectional replication of the *Escherichia coli* chromosome. Nature New Biol 232:137–140

Messer W (1972) Initiation of deoxyribonucleic acid replication in *Escherichia coli* B/r: chronology of events and transcriptional control of initiation. J Bacteriol 112:7–12

Pierucci O, Helmstetter HE, Rickert M, Weinberger M, Leonard AC (1987) Overexpression of the *dnaA* gene in *Escherichia coli* B/r: chromosome and minichromosome replication in the presence of rifampicin. J Bacteriol 169:1871–1877

Pinkel D, Dean P, Lake S, Peters D, Mendelsohn M, Gray J, Van Dilla M, Gledhill B (1979) Flow cytometry of mammalian sperm progress in DNA and morphology measurement. J Histochem Cytochem 27:353–358

Pritchard RH, Zaritsky A (1970) Effect of thymine concentration on the replication velocity of DNA in a thymine-less mutant of *Escherichia coli*. Nature 226:126–131

Salzman GC, Mullaney PF, Price BJ (1979) Light-scattering approaches to cell characterization. In: Melamed MR, Mullaney PF, Mendelsohn ML (eds) Flow cytometry and sorting. John Wiley, New York, pp 105–153

Salzman G, Singham SB, Johnston RG, Bohren CF (1990) Light scattering and cytometry. In: Melamed MR, Lindmo T, Mendelsohn ML (eds) Flow cytometry and sorting, 2nd edn. Wiley-Liss, New York, pp 81–107

Shapiro HM (1988) Practical flow cytometry. Alan R Liss, New York

Skarstad K, Boye E (1988) Perturbed chromosomal replication in *recA* mutants of *Escherichia coli*. J Bacteriol 170:2549–2554

Skarstad K, Steen HB, Boye E (1983) Cell cycle parameters of slowly growing *Escherichia coli* B/r studied by flow cytometry. J Bacteriol 154:656–662

Skarstad K, Steen HB, Boye E (1985) *Escherichia coli* DNA distributions measured by flow cytometry and compared with theoretical computer simulations. J Bacteriol 163:661–668

Skarstad K, Boye E, Steen HB (1986) Timing of initiation of chromosome replication in individual *Escherichia coli* cells. EMBO J 5:1711–1717

Skarstad K, von Meyenburg K, Hansen FG, Boye E (1988) Coordination of chromosome replication initiation in *Escherichia coli*: effects of different *dnaA* alleles. J Bacteriol 170:852–858

Skarstad K, Løbner-Olesen A, Atlung T, von Meyenburg K, Boye E (1989) Initiation of DNA replication in *Escherichia coli* after overproduction of the DnaA protein. Mol Gen Genet 218:50–56

Steen HB (1990a) Flow cytometric studies of microorganisms. In: Melamed MR, Lindmo T, Mendelsohn ML (eds) Flow cytometry and sorting, 2nd edn. Wiley-Liss, New York, pp 605–622

Steen HB (1990b) Characteristics of flow cytometers. In: Melamed MR, Lindmo T, Mendelsohn ML (eds) Flow cytometry and sorting, 2nd edn. Wiley-Liss, New York, pp 11–25

Steen HB (1990c) Light scattering in an arc lamp-based flow cytometer. Cytometry 11:223–230

Steen HB (1991) Flow cytometry instrumentation. In: Demers S (ed) Particle analysis in oceanography, NATO ASI series, vol. G 27, Springer, Berlin, pp 3–29

Steen HB (1992) Noise, sensitivity and resolution in flow cytometers. Cytometry 13 (in press)

Steen HB, Lindmo T, Stokke T (1989) Differential light scattering in an arc lamp-based flow cytometer. In: Yen A (ed) Flow cytometry: advanced research and clinical applications, vol. 1. CRC Press, Boca Baton, pp 63–80

Steen HB, Skarstad K, Boye E (1990) DNA measurements of bacteria. In: Darzynkiewicz Z, Crissman H (eds) Methods in cell biology, vol. 33. Academic Press, San Diego, pp 519–526

van den Engli G, Farmer C (1992) Photobleaching and photon saturation in flow cytometry. Cytometry 13:669–677

von Meyenburg K, Hansen FG (1987) Regulation of chromosome replication. In: Neidhardt FC, Ingraham JL, Low KB, Magasanik B, Schaechter M, Umbarger HE (eds) *Escherichia coli* and *Salmonella typhimurium*: cellular and molecular biology. American Society for Microbiology, Washington, DC, pp 1555–1577

Xu YC, Bremer H (1988) Chromosome replication in *Escherichia coli* induced by oversupply of DnaA. Mol Gen Genet 211:138–142

Chapter 3

Flow Cytometric Analysis of Heterogeneous Bacterial Populations

Richard Allman, Richard Manchee and David Lloyd

Introduction

Historical Perspectives

There is currently much interest in developing instrumental techniques for rapid microbiological analyses, in particular bacterial identification. This stems largely from the limitations of the classical identification procedures that are still the main approach to identifying bacteria today.

If we accept that microbiology as a "science" began with the publication of Robert Hooke's *Micrographia* in 1665, and bacteriology as beginning in 1683 with Anton van Leeuwenhoek's observations of bacteria, then the intervening three centuries have seen remarkably few advances in the methods we use to identify bacteria (Table 3.1). The publication by Cowan and Steel (1965) still provides the main methodologies for identifying bacteria in clinical laboratories today. Thus, clinical and environmental laboratories rely upon a preliminary microscopical examination of the organism, followed by a series of biochemical tests; these often include the ability of the organism to grow under various nutritional and physical conditions, and its sensitivity to antibiotics and bacteriophages. Such an approach takes several days and often yields results that may be ambiguous, difficult to interpret, and produce only a presumptive identification. The newer methodologies using monoclonal antibodies and DNA hybridization probes are, in general, only used as confirmatory tests in identification protocols. Their high specificity is indeed a positive disadvantage when it comes to identifying unknown bacteria, as a separate probe is required for each organism, thus making these techniques very expensive. Therefore, a simple automated technique yielding unambiguous identification of as wide a range of organisms as possible, within minutes, is highly desirable.

Table 3.1. Some landmarks in the history of diagnostic microbiology

1590	Hans and Zacharias Janssen's microscope
1610	Galileo's microscope
1665	Hooke's *Micrographia*
1680	Leeuwenhoek's observations of yeast cells
1683	Leeuwenhoek's observations of bacteria, *Giardia*, etc.
1757	Linnaeus' classification. "Infusoria", including bacteria, were placed in a single genus: "Chaos"
1857–85	Louis Pasteur works on yeasts and bacteria
1876	Aetiology of anthrax (Koch)
1884	Gram's stain
1990	Selective media, e.g. MacConkey for enteric bacteria, Conradi and Drigalski's crystal violet/litmus media for *Salmonella typhi*, Bordet and Gengous' blood potato medium for *Bordetella pertussis*
1923	Bergey's *Manual of determinative bacteriology*, 1st edition
1957	The application of computers to taxonomy (Sneath 1957)
1965	Cowan and Steel's *Manual for the identification of medical bacteria*
1975	Monoclonal antibodies (Kohler and Milstein 1975)
1975	Specific gene probes (Southern 1975)

Table 3.2. Instrumental techniques that have been applied to the problem of identifying microorganisms

Technique	References
Pyrolysis gas chromatography	Stack and Donohue (1978), French et al. (1980), Irwin (1982)
Capillary gas chromatography	Gutteridge and Norris (1979), Drucker (1981)
Capillary gas chromatography/mass spectrometry	Fox and Morgan (1985)
Particle analysis mass spectrometry	Sinha (1985)
Primary fluorescence spectroscopy	Shelly et al. (1980a,b), Rossi and Warner (1985)
Immunofluorescence spectroscopy	Rossi and Warner (1985)
Bioluminescence	Neufeld et al. (1985)
Chemiluminescence	Neufeld et al. (1985)
Raman spectroscopy	Carey (1982), Tu (1982), Hartman and Thomas (1985)
Electrical impedance	Hadley and Yajko (1985)

Instrumental Methods for Rapid Microbiological Analyses

The above considerations have prompted a number of studies into the use of modern instrumental techniques as alternatives to the lengthy "traditional" procedures (Table 3.2). Microchip technology has revolutionized modern scientific instrumentation, which is now characterized by a high degree of automation, high-speed data acquisition, and computer-aided analysis of the data.

Although the techniques listed in Table 3.2 have all proved to be useful for certain particular studies (reviewed in Nelson 1985), they all have drawbacks when applied to the problem of bacterial identification. The gas chromatography and mass spectrometric methods both provide very detailed

Introduction

analysis of the chemical composition of organisms. However, whilst different types of bacteria may appear to be different morphologically, they are in fact remarkably similar at the molecular level, making the analysis of such data extremely difficult. In general, the other techniques listed suffer from the disadvantage that although they can inform us whether or not we have bacteria present in a sample, they are unable to tell us which one(s).

Recent improvements in the sensitivity and specificity of flow cytometric instrumentation and techniques have made possible a wide range of microbiological investigations that have led us to believe that they may offer the necessary techniques for the rapid characterization of bacterial populations.

Flow Cytometry in Microbiology

Flow cytometers as used today were developed from instruments designed to automate the counting of red blood cells. The first device specifically designed for the automatic counting of cells in flow is usually attributed to Moldavan (1934), with little further work appearing until the first commercially successful blood counting instrument was introduced (Coulter 1956). Flow cytometry as we know it today probably owes most to the work of Kamentsky and colleagues (1965, 1967) who introduced two new concepts that greatly extended the potential applications of flow cytometry. The first was the use of spectrophotometry to quantitate specific cellular constituents, and the second was that of cell classification by a combination of multiple simultaneous measurements of different cellular parameters.

Until recently remarkably little work had been reported on flow cytometry in microbiology, despite the lack of suitable alternative techniques (reviewed in Boye and Løbner-Oleson 1990). Paradoxically, one of the first reported applications of flow cytometry was in bacteriology (Ferry et al. 1949).

The lack of progress in microbiological applications of flow cytometry was probably the result of several experimental problems that are encountered when analysing bacteria.

The DNA content of bacteria is typically 3 orders of magnitude lower than that of mammalian cells. Thus measurement of bacterial DNA content with any degree of precision requires techniques that are more sophisticated than those normally used for mammalian cells. The pioneering work of Steen and coworkers (Steen and Lindmo 1979; Steen and Boye 1980, 1981; Steen 1983) showed the necessity for using dyes with a high affinity for DNA and high quantum yield, as well as having an instrument of sufficient sensitivity. Many (all?) of the early commercial laser-based instruments lacked this sensitivity.

Further difficulties are encountered when light scattering measurements are used. The volume of bacteria is also typically 3 orders of magnitude lower than that of mammalian cells. In several commercial laser-based instruments the wide-angle scatter detection, in particular, lacks sufficient sensitivity for bacterial measurements.

The two problems cited above are technical problems that have been solved in several instruments; however, the main experimental difficulty in analysing bacteria is that many of their biological characteristics (including

Table 3.3. Previous reports of flow cytometry in clinical and environmental microbiology

Organism	Source	References
Bacillus anthracis	Pure culture	Phillips and Martin (1983, 1985, 1988)
Escherichia coli	Blood	Mansour et al. (1985)
Corynebacterium spp.	Urine	Van Dilla et al. (1983)
Escherichia coli	Urine	Van Dilla et al. (1983)
Staphylococcus epidermidis	Urine	Van Dilla et al. (1983)
Staphylococcus saprophyticum	Urine	Van Dilla et al. (1983)
Streptococcus faecalis	Urine	Van Dilla et al. (1983)
Legionella pneumophila	Water cooling towers	Ingram et al. (1982)
Legionella spp.	Pure culture	Tyndall et al. (1985)
Listeria monocytogenes	Milk	Donnelly and Baigent (1986)
Naegleria fowleri	River water	Muldrow et al. (1982)
Plasmodium vinckei	Blood	Jacobberger et al. (1983)
Plasmodium berghei	Blood	Howard et al. (1979)
Plasmodium falciparum	Blood	Howard et al. (1979), Whaun et al. (1983)

size, shape and DNA content) vary depending upon the growth conditions used, or the source from which the organisms were obtained. Therefore, strict reproducibility of conditions is required in order to produce consistent data.

There have been several attempts at using flow cytometry to characterize clinically or environmentally important microorganisms (Table 3.3). Previous attempts to demonstrate the use of flow cytometry for bacterial characterization have relied upon immunofluorescence-based methods (Ingram et al. 1982; Phillips and Martin 1983; Tyndall et al. 1985; Phillips and Martin 1985, 1988; Donnelly and Baigent 1986), or involved computerized analyses of measurements obtained with expensive dual-laser instruments (Van Dilla et al. 1983; Sanders et al. 1990). Both of these approaches have major limitations. For routine use in the characterization of heterogeneous bacterial populations, the methodologies need to be simple and rapid, whilst the instrumentation needs to combine sensitivity with simplicity and low cost. The methods chosen must also be applicable to as wide a range as possible of different organisms. The work presented here illustrates the difficulties encountered in such an approach, and possibly points to some solutions in the search for a simpler flow cytometric technique for the rapid characterization of bacteria.

Experimental Approaches

Light Scattering Approaches to Bacterial Characterization

Light scattering measurements are simply and rapidly obtained (requiring no sample preparation or reagent additions) and are, therefore, ideal components of a rapid diagnostic protocol.

Most flow cytometers have the ability to make dual parameter measurements of light scattered by particles at two different angles, i.e. forward ($<10°$) and wide-angle scatter ($>15°$).

Experimental evidence indicates that forward angle light scatter provides a good measure of size for both ideal particles, i.e. latex spheres (Sharpless et al. 1977; Salzman et al. 1979) and cells (Mullaney et al. 1969; Bekkum et al. 1971; Arndt-Jovin and Jovin 1974; Meyer and Brunsting 1975; Loken et al. 1976; Kerker et al. 1979; Salzman et al. 1979) and that wide-angle scatter is a measure of a combination of cell structure (Meyer and Brunsting 1975; Salzman et al. 1975; Visser et al. 1978) and size (Kerker et al. 1979; Salzman et al. 1979). However, light scattering is sensitive to fluctuation in many variables in addition to cell size and shape.

The forward scatter signal is a complex parameter; interpretation is hazardous in that this function varies not only with cell size, but also with cell shape, refractive index and number of intracellular dielectric interfaces (e.g. within the multilayers of the cell envelope, or between cytosol and DNA). Thus any variation in the forward scatter signal may not be directly attributable to a change in cell size, but also may have varying contributions from these other potentially variable characteristics of the organism. Scattering theory also predicts that the relationship between forward scattering amplitude and particle size will not be monotonic; experiments with particles of known sizes confirm this prediction (Salzman et al. 1979). Thus the assumption that forward scatter signals produced by intact cells of a single

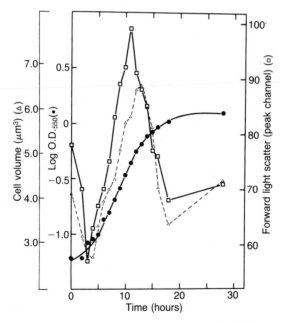

Fig. 3.1. Fluctuation in Coulter cell volume ($\triangle$) and narrow forward angle light scattering (peak channel) ($\square$) during the batch growth of *Azotobacter vinelandii* NCIB 8789. Growth was measured as increase in OD_{550} ($\bullet$). The mean cell volume fluctuated between 2.7 μm (at $T = $ 4 h) and 6.6 μm (at $T = $ 13 h). The forward scatter signals were measured using an Ortho Cytofluorograf 50H equipped with an argon ion laser tuned at 488 nm. The forward scatter signals showed good correlation with the cell volume measurements. (Reproduced with permission from Allman et al. (1990).)

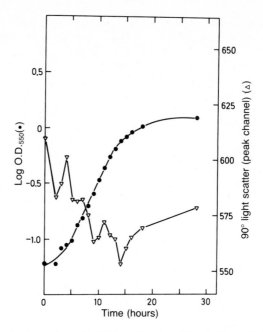

Fig. 3.2. Fluctuation in the wide-angle (90°) scatter signal (peak channel) (▽) during batch growth of *A. vinelandii* NCIB 8789. Growth was measured as increase in OD_{550} (●). The reduction in 90° scatter during mid-exponential growth implies a reduction in refractive index of the cells and hence a reduction in "cellular granularity". (Reproduced with permission from Allman et al. (1990).)

cell type in any given instrument are proportional to cell size may not be completely correct.

The processing of light scatter signals to obtain pulse-width information (time of flight) can yield more accurate estimates of cell size (Sharpless et al. 1975, 1977; Sharpless and Melamed 1976; Leary et al. 1979); however, such measurements are not feasible for bacteria, whose larger dimension is typically less than the width of the incident laser beam.

Light scattered and collected orthogonally to the illumination source (so-called 90°, right-angle or wide-angle scatter) is also a complex parameter. This signal is thought to indicate variations in cell surface structure or

Fig. 3.3a–c. Ultrathin sections of *A. vinelandii* NCIB 8789. Samples were taken 0h (**a**), 13h (**b**) and 28h (**c**) following inoculation (corresponding to lag, mid-exponential and stationary phases of growth). Samples taken after 0 and 28h show the presence of large granular inclusions composed of poly-β-hydroxybutyrate, which is a cellular reserve material for *A. vinelandii*. Both of these samples correspond to high wide-angle scatter signals. The sample taken where growth rate is maximal does not show the presence of these inclusions, and the sample corresponds to a low wide-angle scatter signal. The presence of these inclusions is likely to cause fluctuation in the overall refractive indices of cells, so they are likely to be a major component of the fluctuation in wide-angle scatter signals during the batch growth of *A. vinelandii*. ×23 000

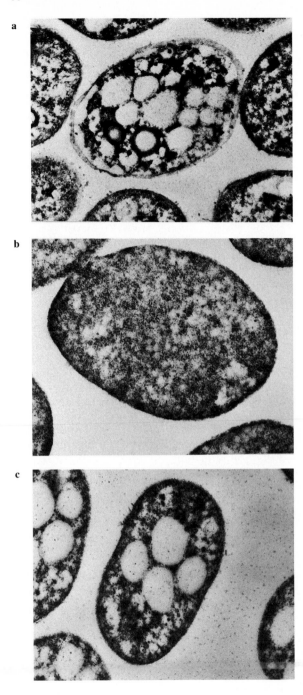

internal structures, which are generally referred to in the literature as "cellular granularity". For particles in the size range of biological cells (0.5–20 μm), the amplitude of the wide-angle scatter signal is approximately 3 orders of magnitude lower than that of the forward scatter signals. This difference in amplitude of the forward and wide-angle scatter signals influences the biological properties that they measure. In the forward direction, scattering from large particles is many orders of magnitude larger than that from small particles, so the whole cross section dominates over any contribution from small particles within the cell, and the signal increases with size. However, at larger angles the scattered light intensity from small particles competes on a more favourable basis with the scattered light intensity from the whole cell cross section, thus allowing differences in cell morphology to show up in the signal.

It has been very elegantly demonstrated that the combination of forward and wide-angle scatter can be used to discriminate a wide range of mammalian cells, in particular haemopoietic cells (Visser et al. 1978, 1980). However, before such measurements are used for bacteria it must be restated that properties such as cell size and shape can vary greatly depending upon the growth conditions. This situation is further complicated due to the fact that cell morphology also varies depending upon the phase of growth at which cells are sampled (Allman et al. 1990).

If we observe the batch culture of the bacterium *Azotobacter vinelandii* NCIB 8789, making measurements at hourly intervals, we see how the forward scatter signal (Fig. 3.1) and the wide-angle scatter signal (Fig. 3.2) vary with growth phase. In this particular case the mean cell volume (as measured using a Coulter counter) fluctuated between 2.7 and 6.6 μm^3, with a corresponding fluctuation in the forward scatter signal. However, the wide-angle scatter profile shows an almost inverse relationship to the forward scatter profile. In order to clarify this apparent anomaly we examined the samples used in the above experiment by electron microscopy. The results confirmed that the reduction in the wide-angle scatter signal for mid-exponential phase cells was caused by structural changes in the organisms (Fig. 3.3). The structural changes involved both internal and surface structures, though it was not possible to say what contribution each of these factors makes to the wide-angle scatter signal. These results illustrate that if we wish to use light scattering measurements in a protocol for characterizing bacteria, we must apply a very strict sampling regime in order that our measurements are reproducible.

It may be argued that the scattering characteristics of bacteria should be well defined mathematically. Cross and Latimer (9) have produced a theoretical treatment of the scattering from *Escherichia coli* cells based on the assumption that *E. coli* is elliptical in shape and composed of just two regions of differing refractive index: the cytoplasm and the cell envelope. This approach appeared to show good agreement with experimental observations. However, a theoretical treatment such as that produced for *E. coli* is unlikely to hold for organisms such as *A. vinelandii* that have a complex and variable internal structure (such variations depending upon the stage of growth of the culture). For these reasons, the wide-angle scatter signal is likely to remain a qualitative parameter in flow cytometric measurements of bacteria. As long as we remain aware of these potential problems,

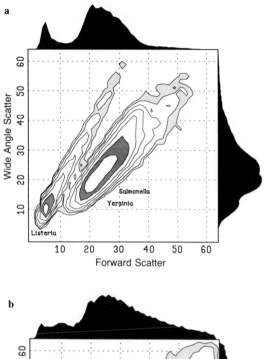

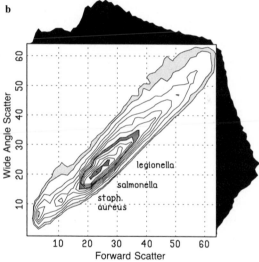

Fig. 3.4a–d. Dual parameter contour plots of forward scatter versus wide-angle scatter for 10 different bacterial cell types. In each case *Salmonella typhimurium* was used as a "reference organism". *Bacillus cereus* and *Clostridium perfringens* showed much larger scattering signals than the other organisms and these are shown on a contour plot in which the photomultiplier gain was reduced by a factor of 2 in order to bring the clusters on scale. Photomultiplier settings were constant for all other samples. (*Continued overleaf.*)

then dual parameter light scatter measurements can indeed provide a useful first step for characterizing bacteria.

We have used a commercially available arc-lamp-based flow cytometer to analyse the light scattering profiles of a number of clinically relevant micro-

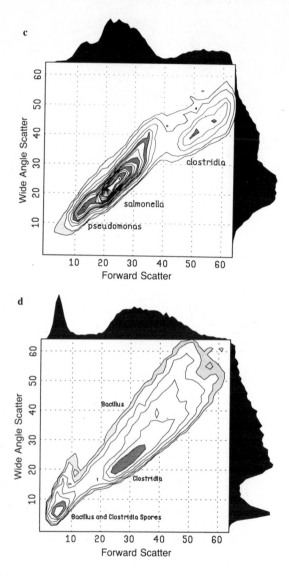

Fig. 3.4. *Continued*

organisms (*Bacillus cereus*, *B. cereus* spores, *Clostridium perfringens*, *C. perfringens* spores, *Legionella pneumophila*, *Listeria monocytogenes*, *Pseudomonas fluorescens*, *Salmonella typhimurium*, *Staphylococcus aureus*, *Yersinia enterocolitica*). Figure 3.4 shows dual parameter contour plots of forward scatter versus wide-angle scatter for mid-exponential phase cells in steady-state growth, together with the two gram-positive spore types. It can be seen that some of the organisms can be distinguished purely on the basis of their differing light scatter properties. However, it is also apparent

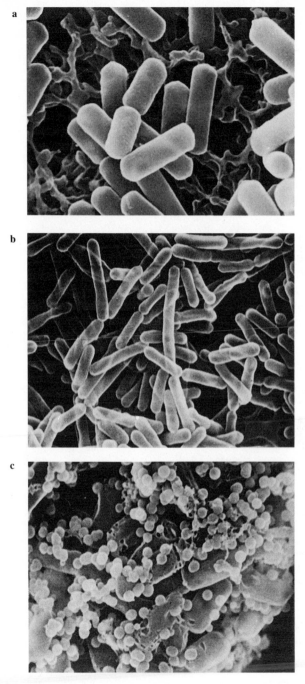

Fig. 3.5a–c. Scanning electron micrographs of *Salmonella typhimurium* (**a**), *Listeria monocytogenes* (**b**) and *Bacillus cereus* spores (**c**). ×12 500

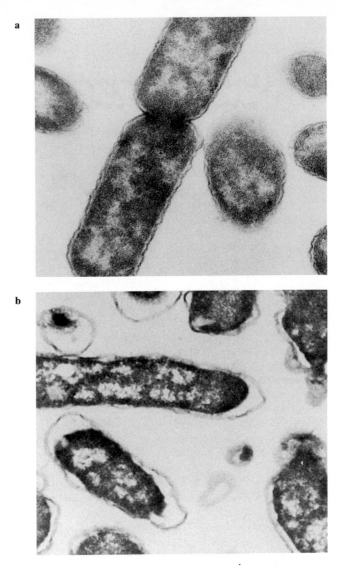

Fig. 3.6a,b. Ultrathin sections of *Salmonella typhimurium* (**a**) and *Legionella pneumophila* (**b**). The differences in internal and surface structure of the organisms are clearly evident. ×28 000

that some of the organisms have very similar (overlapping) scattering characteristics.

We have used scanning and transmission electron microscopy to obtain measurements of size and shape as well as characterization of surface and internal structures to determine which of these factors are contributing to the light scatter profiles. These observations confirm that the generally held assumption that the intensity of forward light scatter is correlated with cell size does not necessarily hold for mixed populations. Thus, *L*.

monocytogenes, which is longer than *S. typhimurium*, gives a relatively small forward scatter signal. That spores of *B. cereus* and *C. perfringens* give a forward scatter signal that is out of proportion to their size, may be explained on the basis of a high value for their refractive index (Fig. 3.5). Analysis of the structural characteristics of the organisms also reveals some anomalies. Thus, *L. pneumophila* and *S. typhimurium* are of similar size and shape, but their internal structures are markedly different (Fig. 3.6). On this basis we would expect them to give rise to distinctive wide-angle scattering signals. In this case, this is clearly not so, and we must conclude that overall "effective" refractive indices of the organisms are similar.

Bacterial DNA Content

The DNA content of the *E. coli* chromosome is approximately 1400 times less than that of a diploid human cell. This indicates that the constraints on the staining protocol and the sensitivity of the instrument are much more stringent than for human cells. The attributes that are important in the choice of fluorescent probe are summarized in Table 3.4.

The method we have found to give the best results in bacteria (i.e. large fluorescent signal, reproducible results) is that developed by Steen and Boye (1980, 1981) and described in detail in Steen et al. (1990). Basically the method involves staining the bacteria with a combination of mithramycin and ethidium bromide. Mithramycin is used as it is DNA-specific (binding to GC-rich regions) and thus obviates the requirements to remove cellular RNA with RNAse treatment. Ethidium bromide is used to increase the fluorescence signal. Mithramycin has a relatively low quantum yield; however, when energy is transferred to ethidium bromide molecules which are close enough (by resonance energy transfer) the net result is an increase in fluorescence intensity by a factor of approximately 2 (Langlois and Jensen 1979).

The small size and low DNA content of bacteria imposes the need for an instrument of sufficient sensitivity. Our laser-based instrument, an Ortho Cytofluorograf 50-H, was found to measure bacterial light scattering adequately but the fluorescence detection was inadequate for measuring bacterial DNA content. The instrument used for the following experiments was a Skatron Argus (Skatron, Norway), which is an arc-lamp-based instrument after the design of Steen and Lindmo (1979; Steen 1983). How-

Table 3.4. Attributes of an ideal fluorescent DNA probe

1. High specificity for DNA binding (many DNA stains also bind to RNA and other cell components)
2. The dye should have access to frequent binding sites on the DNA molecule (i.e. as many sites as possible per phosphate residue)
3. The dye should have a high extinction coefficient (i.e. absorb as many photons as possible)
4. The dye should have a high quantum yield (The maximum possible value of the fluorescence quantum yield is 1, whereby, for every photon absorbed, a photon is emitted as fluorescence. In practice values close to 1 are rarely achievable)
5. The dye should not be sensitive to DNA–protein interactions

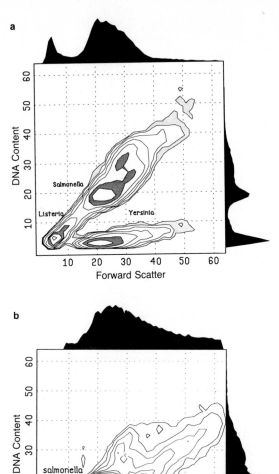

Fig. 3.7a–d. Dual parameter (forward scatter versus DNA content) contour plots of the 10 bacterial cell types. The use of DNA content as one of the measurement parameters provides much improved resolution of the organisms. Bacterial DNA was stained with a combination of mithramycin (Pfizer; 90 μg/ml) and ethidium bromide (Sigma; 25 μg/ml) in 10 mmol/l tris HCl, pH 7.6, following fixation in ice-cold 70% (v/v) ethanol (Steen et al. 1990).

ever, several of the newer laser-based instruments have improved optical designs and are able to make adequate measurements on bacteria.

Figure 3.7 illustrates how simple dual parameter measurements of forward scatter and DNA content can be used to further resolve populations of different bacteria. The *Listeria* and spore populations still overlap to some

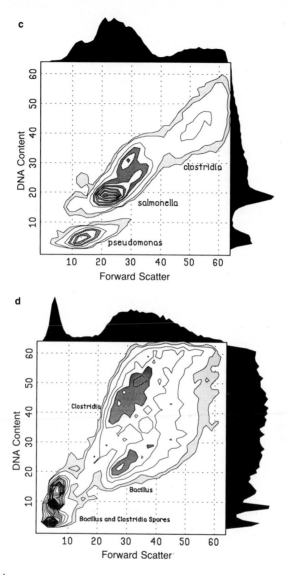

Fig. 3.7. *Continued*

extent and it was not possible to distinguish the two spore populations (*B. cereus* and *C. perfringens*) under these conditions.

Although this is a very simple example with just a few organisms it does show that, in principle, bacteria can be characterized with only a few simple analyses. However DNA/light scatter measurements *per se* cannot provide a definitive identification of bacteria, although it is reasonably likely that a suitable combination of parameters can be found that when used in parallel with DNA/light scatter measurements will unequivocally distinguish one bacterial type from another.

Functional Probes in Bacterial Characterization

Cellular parameters measurable by flow cytometry may be described as being "structural" or "functional" (Shapiro 1983). Thus, forward light scatter, wide-angle light scatter and DNA content are structural measurements. However, there are a number of functional parameters, i.e. measurements of biological activities, which may be useful in characterizing bacteria. Such measurements include enzyme activity, intracellular pH, surface charge, surface receptors, cytoplasmic calcium and membrane potential. Of these possibilities, the use of membrane potential probes holds many attractions.

Several investigators have used simple electronic particle counters to detect bacteria in urine (Dow et al. 1979; Alexander et al. 1981). However, such measurements are meaningless in most instances, since the presence of particles is not necessarily indicative of bacteria, and if bacteria are present, such measurements give us no clue as to the identity of those organisms.

Paradoxically those classical measurements that were described in the introduction as being time-consuming (i.e. growth on different media) may find a place in flow cytometric methods for bacterial identification using membrane potential probes.

The measurement of bacterial membrane potential (in terms of defining precise membrane potential voltages) using optical probes is difficult (Shapiro 1990), and there is still debate regarding the mechanism of dye

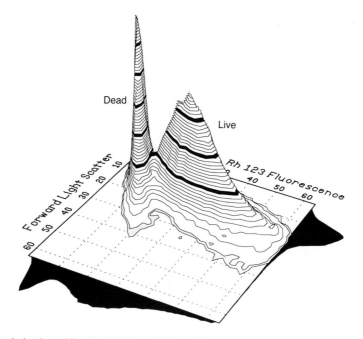

Fig. 3.8. Discrimination of live and dead populations of *Staphylococcus aureus*. The organisms were stained with rhodamine 123 (Sigma; 0.5 μg/ml for 15 min) in the growth medium (Nutrient Broth, Difco). Dead cells were obtained by heating 5 ml of cell suspension to 60 °C for 15 min prior to staining.

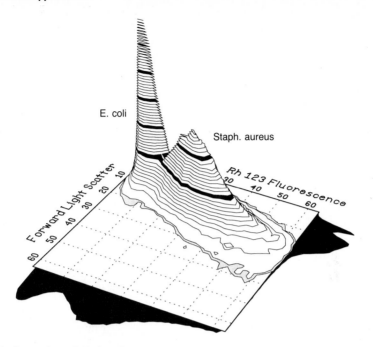

Fig. 3.9. Separation of *Escherichia coli* and *Staphylococcus aureus* following staining with rhodamine 123 (Sigma; 0.5 µg/ml) for 15 min in the growth medium (Nutrient Broth, Difco). The reduced uptake of the dye by *E. coli* is due to the outer membrane of this organism.

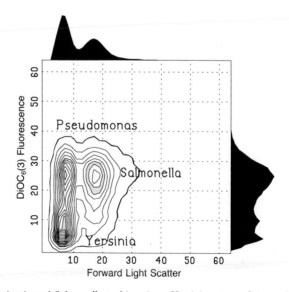

Fig. 3.10. Discrimination of *Salmonella typhimurium*, *Yersinia enterocolitica* and *Pseudomonas fluorescens* on the basis of dual parameter measurements of forward light scatter and cyanine dye fluorescence. Organisms were stained with the cyanine dye 3,3-dihexyloxacarbocyanine iodide (Sigma; 5 µg/ml for 10 min) in the growth medium (Nutrient Broth, Difco).

uptake in response to membrane potential. However, in terms of bacterial identification such dyes can give much information. In the first instance, any particle in the size range of bacteria that is found to have a membrane potential is therefore identified as a viable microorganism (dead cells are depolarized and have no membrane potential) (Fig. 3.8).

A further important piece of diagnostic information available from the use of such probes depends upon the fact that the outer membrane of gram-negative organisms renders them much less permeable to the probes than gram-positive organisms, thus providing a "high-tech Gram stain" (Fig. 3.9). Gram-positive/negative discrimination would be a very desirable addition to any diagnostic protocol. However, on top of all these advantages it turns out that different organisms show different distributions of dye uptake (presumably due to them having different membrane potentials) (Fig. 3.10). Experiments by Shapiro (1988) suggest that measurements of bacterial membrane potential provide separations that are equally as good as those provided by measurements of GC/AT ratios.

Neural Networks in Flow Cytometric Data Processing

Now that we have some insight into the biological characteristics that are important for the flow cytometric characterization of bacteria, the remaining problem is one of analysing the data. Flow cytometry provides a lot of data in a very short time, thus any automation of the identification of bacteria (or any other cell types) from such measurements requires computer software that can also analyse the data very quickly. We have now demonstrated that back-propagation neural networks provide an excellent tool for doing this job (Morris et al. 1992).

The general principles of neural networks are described in numerous texts (e.g. Simpson 1990). In short they provide a means of "teaching" a computer to "recognize" complicated data patterns such as the flow cytometric profiles of different bacteria or other cells/organisms. Thus when the bacterium is encountered in a real (unknown) sample, the computer is able to identify the sample almost in real time!

Discussion

Rapid automated cell identification in haematology and cancer cytology provided the driving force behind the development of flow cytometry. It is, therefore, appropriate that with the increasing application of flow cytometry in microbiology, the problem of bacterial classification is providing a new challenge. Flow cytometry is now the standard technique for classifying lymphocytes. It may be that flow cytometry will occupy an equally prominent position in microbiology in the near future.

There is still a need for flow cytometric instrumentation to be improved further for routine microbiological use, in particular with regard to ease of operation and financial cost of the equipment. In the field of organic chemistry we also need newer dyes with higher quantum yields thus reducing the demands upon (and possibly the cost of) flow cytometric instrumentation.

Acknowledgement

The authors wish to thank The Ministry of Defence for financial support and Dr. Ao C. Hann for expert help with electron microscopy.

References

Alexander MS, Khan MS, Dow CS (1981) Rapid screening for bacteriuria using a particle counter, pulse height analyser, and computer. J Clin Pathol 34:194–198

Allman R, Hann AC, Phillips AP, Martin KL, Lloyd D (1990) Growth of *Azotobacter vinelandii* with correlation of coulter cell size, flow cytometric parameters, and ultrastructure. Cytometry 11:822–831

Arndt-Jovin DJ, Jovin TM (1974) Computer controlled multiparameter analysis and sorting of cells and particles. J Histochem Cytochem 22:622–625

Bekkum DWU, Noord MJV, Maat A, Dicke KA (1971) Attempts at identification of haemopoietic stem cells in mouse. Blood 38:547–558

Boye E, Løbner-Olesen A (1990) Flow cytometry: illuminating microbiology. New Biologist 2:119–125

Carey PR (1982) Biological applications of raman and resonance raman spectroscopy. Academic Press, New York

Cowan ST, Steel KJ (1965) Manual for the identification of medical bacteria. Cambridge University Press, Cambridge

Coulter WH (1956) High speed automatic blood cell counter and cell size analyser. Proc Natl Electronics Conf 12:1034

Cross DA, Latimer P (1972) Angular dependence of scattering from *Escherichia coli* cells Appl Optics 11:1225–1228

Donnelly CW, Baigent GJ (1986) Method for flow cytometric detection of *Listeria monocytogenes* in milk. Appl Environ Microbiol 52:689–695

Dow CS, France AD, Khan MS, Johnson T (1979) Particle size distribution analysis for the rapid detection of microbial infection in urine. J Clin Pathol 32:386–389

Drucker DB (1981) Microbiological applications of gas chromatography. Cambridge University Press, Cambridge

Ferry RM, Farr LE, Hartman MG (1949) The preparation and measurement of the concentration of dilute bacterial aerosols. Chem Rev 44:389–395

Fox A, Morgan SL (1985) The chemotaxonomic characterisation of microorganisms by capillary gas chromatography and gas chromatography mass spectrometry. In: Nelson WH (ed) Instrumental methods for rapid microbiological analysis. VCH Inc., USA, pp 135–162

French GL, Gutteridge CS, Phillips I (1980) Pyrolysis gas chromatography of *Pseudomonas* and *Acinetobacter* species. J Appl Bacteriol 49:505–516

Gutteridge CS, Norris JR (1979) The application of pyrolysis techniques to the identification of microorganisms. J Appl Bacteriol 47:5–43

Hadley WK, Yajko DM (1985) Detection of microorganisms and their metabolism by measurements of electrical impedance. In: Nelson WH (ed) Instrumental methods for rapid microbiological analysis. VCH Inc., USA, pp 193–209

Hartman KA, Thomas GJ (1985) The identification, interactions and structure of viruses by raman spectroscopy. In: Nelson WH (ed) Instrumental methods for rapid microbiological analysis. VCH Inc., USA, pp 91–134

Howard RJ, Battye FL, Mitchell GF (1979) Plasmodium infected blood cells analysed and sorted by flow cytometry with deoxyribonucleic acid binding dye 33258 Hoechst. J Histochem Cytochem 27:803–813

Ingram M, Cleary TJ, Price BJ, Price RL, Castro A (1982) Rapid detection of *Legionella pneumophila* by flow cytometry. Cytometry 3:134–137

Irwin WJ (1982) Analytical pyrolysis: A comprehensive guide. Marcel Dekker, New York, pp 381–431

Jacobberger JW, Horan PK, Hare JD (1983) Analysis of malaria parasite infected blood by flow cytometry. Cytometry 4:228–234

Kamentsky LA, Melamed MR, Derman H (1965) Spectrophotometer: a new instrument for ultrarapid cell analysis. Science 150:630–631

Kamentsky LA, Melamed MR (1967) Spectrophotometric cell sorter. Science 156:1364–1365

Kerker M, Chew H, McNulthy PJ et al. (1979) Light scattering and fluorescence by small particles having internal structure J Histochem Cytochem 27:250–263

Kohler G, Milstein C (1975) Continuous cultures of fused cells secreting antibody of predefined specificity. Nature 256:495–497

Langlois RG, Jensen RH (1979) Interactions between pairs of DNA specific fluorescent stains bound to mammalian cells. J Histochem Cytochem 27:72–78

Leary JF, Todd P, Wood JCS, Jett JH (1979) Laser flow cytometric light scatter and fluorescence pulse width and pulse rise time sizing of mammalian cells. J Histochem Cytochem 27:315–320

Loken MR, Sweet RG, Herzenberg LA (1976) Cell discrimination by multiangle light scattering. J Histochem Cytochem 24:284–291

Mansour JD, Robson JA, Arndt GW, Schulte TM (1985) Detection of *E. coli* in blood using flow cytometry. Cytometry 6:186–190

Meyer RA, Brunsting A (1975) Light scattering from nucleated biological cells. Biophys J 15:191–203

Moldavan (1934) A photoelectric technique for the counting of microscopical cells. Science 80:188–189

Morris CW, Boddy L, Allman R (1992) Identification of basidiomycete spores by neural network analysis of flow cytometry data. Mycological Res 96:697–701

Muldrow LL, Tyndall RL, Fliermans CB (1982) Application of flow cytometry to studies of pathogenic free living amoebae. Appl Environ Microbiol 44:1258–1269

Mullaney PE, Van Dilla MA, Coulter JR, Dean PN (1969) Cell sizing: a light scattering photometer for rapid volume determination. Rev Sci Instruments 40:1029–1032

Nelson WH (1985) Instrumental methods for rapid microbiological analysis. VCH Inc., USA

Neufeld HA, Pace JG, Hutchinson RW (1985) Detection of microorganisms by bio- and chemiluminescence techniques. In: Nelson WH (ed) Instrumental methods for rapid microbiological analysis. VCH Inc., USA, pp 51–65

Phillips AP, Martin KL (1983) Immunofluorescence analysis of *Bacillus* spores and vegetative cells by flow cytometry. Cytometry 4:124–129

Phillips AP, Martin KL (1985) Dual parameter scatter-flow immunofluorescence analysis of *Bacillus* spores. Cytometry 6:124–129

Phillips AP, Martin KL (1988) Limitations of flow cytometry for the specific detection of bacteria in mixed populations. J Immunol Meth 106:109–117

Rossi TM, Warner IM (1985) Bacterial identification using fluorescence spectroscopy. In: Nelson WH (ed) Instrumental methods for rapid microbiological analysis. VCH Inc., USA, pp 1–50

Salzman GC, Crowell JM, Martin JC et al. (1975) Cell classification by laser light scattering: Identification of unstained leukocytes. Acta Cytol 19:374–377

Salzman GC, Wilder ME, Jett JH (1979) Light scattering with stream-in-air flow systems. J Histochem Cytochem 27:264–267

Sanders CA, Yajko DM, Hyun W et al. (1990) Determination of guanine plus cytosine content of bacterial DNA by dual laser flow cytometry. J Gen Microbiol 136:359–365

Shapiro HM (1983) Multistation multiparameter flow cytometry: a critical review and rationale. Cytometry 3:227–243

Shapiro HM (1988) Practical flow cytometry, 2nd edn. Liss-Wiley, New York, pp 297–298

Shapiro HM (1990) Cell membrane potential analysis. In: Darzynkiewicz Z, Crissman HA (eds) Methods in cell biology, vol 33: Flow cytometry. Academic Press, London, pp 25–35

Sharpless TK, Melamed MR (1976) Estimation of cell size from pulse shape in flow cytofluorometry. J Histochem Cytochem 24:257–264

Sharpless TK, Traganos F, Darzynkiewicz Z, Melamed MR (1975) Flow cytometry: discrimination between single cells and cell aggregates by direct size measurements. Acta Cytol 19:577–581

Sharpless TK, Bartholdi M, Melamed MR (1977) Size and refractive index dependence of simple forward angle scattering measurements in a flow system using sharply focused illumination. J Histochem Cytochem 25:845–856

Shelly DC, Quarles JM, Warner IM (1980a) Identification of fluorescent *Pseudomonas* species. Clin Chem 26:1127–1132

Shelly DC, Warner IM, Quarles JM (1980b) Multiparameter approach to the fingerprinting of fluorescent pseudomonads. Clin Chem 26:1419–1424

Simpson PK (1990) Artificial neural systems. Pergamon Press, Oxford

Sinha MP (1985) Analysis of individual biological particles in air. In: Nelson WH (ed) Instrumental methods for rapid microbiological analysis. VCH Inc., USA, pp 165–189

Sneath PHA (1957) The application of computers to taxonomy. J Gen Microbiol 17:201–226

Southern EM (1975) Detection of specific sequences among DNA fragments separated by gel electrophoresis. J Mol Biol 98:503–517

Stack MV, Donohue HD, Tyler JE (1978) Discrimination between oral streptococci by pyrolysis gas liquid chromatography. Appl Environ Microbiol 35:45–50

Steen HB (1983) A microscope based flow cytophotometer. Histochem J 15:147–150

Steen HB, Boye E (1980) Bacterial growth studied by flow cytometry. Cytometry 1:32–36

Steen HB, Boye E (1981) *Escherichia coli* growth studied by dual parameter flow cytophotometery. J Bacteriol 145:1091–1094

Steen HB, Lindmo T (1979) Flow cytometry: a high resolution instrument for everyone. Science 204:403–404

Steen HB, Skarstad K, Boye E (1990) DNA measurements of bacteria. In: Darzynkiewicz Z, Crissman HA (eds) Methods in cell biology, vol. 33: Flow cytometry. Academic Press, London, pp 519–527

Tu AT (1982) Raman spectroscopy in biology. Wiley, New York

Tyndall RL, Hand RE, Mann RC, Evans C, Jernigan R (1985) Application of flow cytometry to detection and characterisation of *Legionella* spp. Appl Environ Microbiol 49:852–857

Van Dilla MA, Langlois RG, Pinkel D, Yajko D, Hadley WK (1983) Bacterial characterisation by flow cytometry. Science 220:620–622

Visser JWM, Cram LS, Martin JC, Salzman GC, Price BJ (1978) Sorting of a murine granulocyte progenator cell by use of laser light scattering measurements. In: Lutz D (ed) Pulse cytophotometry, part 3. European Press, Ghent, pp 187–192

Visser JWM, Engh GJ, Bekkum DW (1980) Light scattering properties of murine haemopoietic cells. Blood Cells 6:391–407

Whaun JM, Rittershaus C, Ip SH (1983) Rapid identification and detection of parasitized human red cells by automated flow cytometry. Cytometry 4:117–122

Chapter 4

On the Determination of the Size of Microbial Cells Using Flow Cytometry

Hazel M. Davey, Chris L. Davey and Douglas B. Kell

Introduction

In most flow cytometers the determination of cell size is based on the measurement of light scattered by cells as they pass through an illumination zone. In conventional instruments the source of this illumination is a laser (Shapiro 1988). Given that both the cell volume and the DNA content of bacteria is some 1000-fold less than that of higher eukaryotic cells, however, laser-based flow cytometers have generally proved unsuitable for the study of microorganisms (Steen et al. 1990). In the Skatron Argus 100 flow cytometer, a high-pressure mercury arc lamp is used as the excitation source, and the machine makes use of an open flow chamber in which a jet impinges at an angle onto the surface of a microscope cover slip (Fig. 4.1). The result is a flat, laminar flow of water across the glass surface. This flow has only two interfaces – the glass/water interface and the water/air interface – and of these only the former can collect particles that may cause background scattering of light (Steen et al. 1989). Furthermore the orientation of these surfaces perpendicular to the optical axis means that the surfaces themselves scatter only the minimum of light. Thus the system has a high signal-to-noise ratio and is therefore ideal for detecting light scattered by microorganisms (Boye et al. 1983).

When non-spherical cells such as erythrocytes or rod-shaped bacteria are carried in a fast-moving liquid (such as the sheath fluid in a flow cytometer) they will tend to align in the direction of their longest dimension, i.e. with their "flat" sides parallel to the flow (Kachel et al. 1990). When such oriented cells interact with a laser beam that is perpendicular to the flow, the scattering will be larger than if the cells were arranged randomly. However, the optical arrangement of the arc-lamp-based flow cytometer results in the cells being illuminated from a wider range of angles and therefore the scattering signal is less sensitive to cell alignment (Steen and Lindmo 1985).

When a particle (e.g. a cell) interacts with a beam of light, such as that

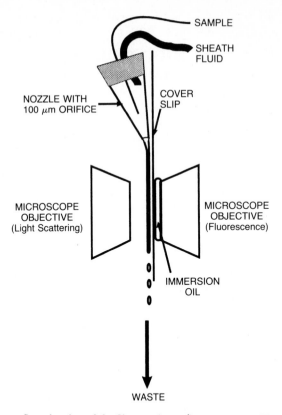

Fig. 4.1. The open flow chamber of the Skatron Argus flow cytometer. The design of this flow chamber causes very little background light scatter, thus giving a high signal-to-noise ratio, making it ideal for measurements on microorganisms.

produced by a mercury arc lamp, some of the light is scattered out of the beam (Mullaney and Dean 1970; McCoy and Lovett 1989; Salzman et al. 1990). The way in which cells scatter light is rather complex and is dependent on their size, shape and internal structure. The relative contributions that these cell characteristics make to the amount of light that is scattered varies with the range of angles over which scattered light is collected, and in particular are functions of the relationship between the cell size and the wavelength of light (Steen and Lindmo 1985; Steen et al. 1989). Generally, the amount of forward scattering (small-angle scattering) increases rapidly with cell size and is affected to a much lesser extent by cell shape and refractive index, while the cell structure and shape become more important at larger angles (Paau et al. 1977; Muirhead et al. 1985; Steen and Lindmo 1985; Wittrup et al. 1988; Boye and Løbner-Olesen 1990; Shapiro 1990). Whilst detection of light scattering at two separate angles has proved useful in the differentiation of blood cell types (Shapiro 1988; Steen et al. 1989), and occasionally for distinguishing different bacteria in mixed populations (e.g. Frelat et al. 1989; Steen 1990), it is usually assumed, after calibration with standards of known sizes, that the extent of low-angle forward light scatter provides an accurate representation of cell size.

It is important to note that flow cytometry yields information on individual cells rather than giving an average value for all of the cells measured, and the special power of the technique is that it allows one to quantify the *heterogeneity* of the sample of interest (Kell et al. 1991). A representation of the cell size distribution can be obtained by the detection of forward scattered light from cells passing in a hydrodynamically focussed stream past a measuring point (Shapiro 1983). Data are collected by appropriate detectors that may be gated electronically, and converted into pulses, the magnitude of which represent the amount of light that is scattered. The pulses are "binned" into channels that increase in number with increasing levels of scattered light (cell size). The data are usually plotted as a histogram in which the abscissa represents the channel numbers whilst the ordinate represents the number of cells measured in each channel (Dean 1990).

While channel numbers allow one to express the data in terms of the *relative* light scattering of each cell, it is also useful for many purposes to be able to express cell sizes in *absolute* terms. This is necessary, for example, if one wishes to determine the mode of growth of individual cells (Davey et al. 1990a). The extent of light scattering can of course be affected by the way in which the flow cytometer is set up (e.g. the flow rate of the sample and sheath fluids, and especially the deterioration of the arc lamp over time (Horan et al. 1990)), and this makes comparisons between work done on different days impossible. By using a cocktail of monodisperse latex beads of known diameters, however, it is possible to obtain a calibration curve of channel number versus cell size, and thus to convert channel numbers to a measure of cell size that is independent of the performance of the flow cytometer. This method is convenient as monodisperse latex beads of high uniformity (coefficient of variation typically <2%) are available in a variety of sizes. However, we have noticed that calibration by this method results in an underestimation of the true cell size, as measured microscopically (Davey et al. 1990a,b) or via the Coulter counter (unpublished observations).

The heterogeneity of cell populations, even in stationary phase, can be quite large (Kell et al. 1991) and therefore it is often desirable to use logarithmic amplification when acquiring flow cytometric data. The latex bead calibration described above also serves to linearize the relationship between the channel number and the diameter of the latex beads.

Often in flow cytometric studies it is a requirement that cells are fixed, either for storage or as a preliminary step in a staining procedure. When cells are stained, for example with a fluorescent DNA stain, measurements are often gated on cell size and the data are plotted as a dual-parameter histogram of DNA versus size versus counts. It is therefore important to determine what effects fixing has on cell size and light scattering behaviour. The purpose of fixing cells prior to staining is to permeabilize the cell membrane, allowing entry of the probe. However, this will also facilitate leakage of cell contents and allow the suspension medium to enter the cell. This may be expected to affect both the "true" cell size and the light scattering properties of the cell as compared both with an unfixed cell and with a latex bead of equivalent size. The effect of a variety of fixatives (ethanol, formalin, etc.) on samples for flow cytometry was investigated by Alanen et al. (1989), who concluded that ethanol fixation produced data for the distribution of DNA content most similar to those obtained with unfixed cells. Many workers (e.g. Kogoma et al. 1985; Skarstad et al. 1985, 1986; Scheper et al. 1987) use 70% ethanol for fixing their cells, although we are not aware of any report describing the effect of this on the apparent cell size.

In view of the above, it seemed important to perform a detailed and systematic study of the effect of fixation and sample preparation on the relationship between the channel numbers observed and the "true" cell size. We describe here the results of just such a study, using strains of *Saccharomyces cerevisiae* and *Micrococcus luteus*. These organisms were chosen because they are almost spherical, since calibration in flow cytometry is usually done with latex spheres. It was anticipated that this choice would help to reduce artifacts due to effects of shape on the (calibrated) data. Since the cells used were all in stationary phase the percentage of (yeast) cells with buds was very low – in all cases 13% or less as judged by scoring 100 cells; thus cell asymmetry due to buds was kept to a minimum.

Materials and Methods

Organisms Used

Three industrial strains of the yeast *Saccharomyces cerevisiae* (BB1, DCL1 and DCL2) were used. BB1 was grown up to high biomass in batch culture in a medium consisting of 1.3% E-broth and 5% malt extract (both from Lab M), pH 6. The working volume of the fermentor was 5 litres, and the temperature was controlled at 30 °C. After 22 h of growth, by which time the culture had reached stationary phase, the contents of the fermentor were harvested by centrifuging the medium from the fermentor to give a pellet. The two other strains of yeast, DCL1 and DCL2, were obtained as a paste; from the low bud counts (less than 13%) these were deemed to have been grown to stationary phase.

Micrococcus luteus (Fleming strain 2665) was grown in batch culture in 5 litres of E-broth (pH 7.4) at 30 °C at pH 7.4. After 34 h the contents of the fermentor were harvested as described for the BB1 yeast above.

Buffers

Phosphate-buffered saline (pH 7.4) was obtained from the Sigma Chemical Company, Poole, Dorset (cat no. P-1000-3). When reconstituted in distilled water it contained (in mmol/l) NaCl 120, KCl 2.7 and phosphate buffer salts 10. This was stored at 4 °C between experiments.

Suspension buffer contained (in mmol/l) KH_2PO_4 50, $MgSO_4$ 5. It was adjusted to pH 7.0 with 5 mol/l KOH. SB was made up freshly for each experiment.

Fluorescein isothiocyanate (FITC) for protein staining was from Sigma. All other reagents were from BDH, Poole, Dorset. Water was singly-distilled in all-glass apparatus.

Preparation of Cells

The pellet of cells, either harvested from a batch culture as described above or cut from a block of paste, was suspended in at least three times its own volume of suspension buffer. The resulting suspension was centrifuged at 1000 g (yeast) or 2750 g (*M. luteus*) for 10 min at room temperature. The supernatant and any precipitated medium from the top of the pellet were discarded and the cells were resuspended in fresh suspension buffer. These were left at room temperature for 45 min, and were occasionally mixed to resuspend them. The cells were recentrifuged as described above and the supernatant discarded. A final wash was performed in suspension buffer followed by another centrifugation to obtain a pellet of cells.

A small amount of the pellet was resuspended in fresh suspension buffer to give approximately 10^7 cells/ml. Two 1 ml aliquots of this suspension were placed in Eppendorf tubes and centrifuged in a bench-top centrifuge at 13 000 rpm. One aliquot was resuspended in suspension buffer, the other was suspended in phosphate-buffered saline. Both were washed twice in their respective buffers before being finally resuspended in suspension buffer or phosphate buffer.

Fixed Cells

The cells in buffer were added to absolute ethanol to give a final ethanol concentration of 70% (v/v). These cells were then allowed to fix at room temperature for 20 min prior to being stored at 4 °C until required. All cells stored in this way were examined within 48 h. Prior to flow cytometry and/or photography of these fixed samples, the cells were centrifuged and resuspended in their original buffers at room temperature.

Cell Treatments Used in the Experiments

Four treatments of cells from each organism were used for flow cytometry and photography. These were:

1. Cells in Phosphate-Buffered saline which were Unfixed (PBU).
2. Cells in Phosphate-Buffered saline which were Fixed (PBF).
3. Cells in Suspension Buffer which were Unfixed (SBU).
4. Cells in Suspension Buffer which were Fixed (SBF).

These cell treatments will be referred to using the three-letter abbreviations shown in parentheses after the descriptions above. The fixed cells were stored as described; the unfixed cells were examined as soon as they were prepared.

Protein Staining

A stock solution of FITC was prepared consisting of 1 mg/ml FITC in acetone; this was stored at 4 °C. On the day of the experiment 50 μl of stock solution was diluted to 2 ml with phosphate buffer (final FITC concentration 25 µg/ml).

Two samples of *Micrococcus luteus* cells in phosphate buffer, one of which had been fixed and the other unfixed, were used for protein staining. The fixed cells were washed twice in phosphate buffer to remove traces of the fixative (ethanol) prior to the addition of 1 ml of staining buffer to the pelleted cells.

Flow Cytometry

Flow cytometry was carried out using a Skatron Argus 100 flow cytometer (Skatron, Ltd, PO Box 34, Newmarket, Suffolk), as described by Steen, Boye and colleagues (Steen et al. 1990; Boye and Løbner-Olesen 1991). The flow cytometer was set up as described in the manufacturer's manual, save that an additional 0.1 µm filter was placed in the sheath fluid line. The sheath fluid was prepared from water that had been filtered using Millipore Milli-Q apparatus. Sodium azide (1 mmol/l) was added to inhibit microbial growth in the sheath fluid tank.

The B1 filter block (excitation, 395–440 nm; band stop, 460 nm; emission, >470 nm) was used for all cell size measurements unless otherwise stated. For experiments in which cells were stained with FITC the FITC filter block was used (excitation, 470–495 nm; band stop, 510 nm; emission, 520–550 nm). The photomultiplier (PMT) setting chosen necessarily differed between the cell types used so that the cell size distribution was on scale. In all cases the gain was logarithmic. A sheath fluid pressure of 1.5–2 kPa/cm^2 was used and the sample flow speed was set at 1 or 2 µl/min. In the experiments described here data were collected from the low-angle light scattering (LS1) detector and, if appropriate, from the FITC fluorescence detector. The flow cytometer was controlled by a Viglen IIHDE micro-computer (IBM-PC-AT-compatible, 80286 processor, EGA screen) with software supplied by the manufacturer. Using this software, data were acquired and saved according to the flow cytometry data file standard out-lined by Murphy and Chused (1984).

Initial calibration and linearization of the cell size measurements were achieved by using a cocktail of monodisperse (cv <2%) latex particles (Dyno Particles AS, PO Box 160, Lillestrøm, Norway). Beads of diameter 2, 5, 7 and 10 µm were run through the flow cytometer using the same settings as for the cells. A three-term polynomial ($y = a + bx + cx^2$) was used to fit a curve of channel number versus cell size to the bead data. The cell data were imported into the spreadsheet program FLOWTOVP.WKS (Davey et al. 1990b) via a filter program written in-house (in Microsoft Quick BASIC v4.5). FLOWTOVP.WKS was used to convert channel numbers to apparent cell diameters using the values fitted to the polynomial.

Photography

Light microscopy was used for all photomicrographs because, although this gives lower resolution than does electron microscopy, it avoids the need for special cell preparation methods that may affect cell size. A Polyvar microscope incorporating a 35 mm camera was used for all photography. Photographs were taken under oil immersion at a total magnification of ×1250. Photographs of the 10 μm beads used in the calibration of the flow cytometer were also used in the calibration of the photomicrographs.

The photographs were developed and printed at a magnification of ×3.5 except for the photographs of *Micrococcus luteus* (and the corresponding set of latex beads), which were magnified to ×10. The resulting photographs for each organism were mixed into random order, identified only by a code number, so that size measurements were carried out blind. The photographs were placed in turn under a ×1.5 bench magnifying glass and the long and short axes of the cells were measured (to the nearest 0.1 mm) using a pair of vernier callipers; once the magnification had been taken into account this corresponded to approximately 0.1 μm. Every cell (apart from unseparated buds) of *Saccharomyces cerevisiae* that was in focus was measured, but in the case of the photographs of *M. luteus*, only the in-focus cells from one randomly selected quarter of the photograph were measured. In most cases 100–200 cells were measured per treatment for each organism. However in some cases, where excessive clumping occurred (see later discussion of this problem), fewer cells were measured. The lowest number of cells measured was 47 and the highest number was 222. All measurements were made by the same person, with the same pair of callipers, to keep systematic errors constant throughout.

The dimensions of the cells (recorded in millimetres from the photographs) were entered into a spreadsheet and converted to their true sizes by using a calibration based on the mode size of a sample of 10 μm beads that had been measured as described above. The length (in micrometres) of the long and short axes of the cells were then used to calculate the diameter of the sphere of equivalent volume.

The volume of an ellipsoid of revolution (Grant et al. 1978) is given by:

$$V = \frac{4}{3} \cdot \pi \cdot a \cdot b^2 \tag{4.1}$$

where V is the volume, a is the long semi-axis and b is the short semi-axis of the ellipsoid. From this one can derive Eq. (4.2) that gives the equivalent diameter for a sphere with the same volume as the ellipsoid:

$$D = 2 \cdot [(L/2) \cdot (S/2)^2]^{\frac{1}{3}} \tag{4.2}$$

where D is the diameter of the equivalent sphere, L is the length of the long axis of the cell and S is the length of the short axis of the cell.

Histograms of the resulting equivalent diameters were produced using in-house software written in Microsoft Quick BASIC v4.5. The program (BIN.BAS/EXE) read in an ASCII list of numbers separated by carriage returns and reported the smallest and largest numbers in the file. The user then entered the required lower and upper bounds (in all cases these were set to 0 μm and 15 μm respectively, which easily encompassed all of the data) and also a suitable "bin" size. Here a bin size interval of 0.2 μm was chosen as, in preliminary tests with a range of bin sizes, it gave good sensitivity without producing an excessively noisy plot. Cells were placed in a given bin if they were equal to or greater than the lower bound and were less than (but not equal to) the upper bound of that particular bin. The output from the BASIC program was in the form of two columns; column one contained the centre value of each bin interval (i.e. 0.1, 0.3, 0.5, ..., 14.9) and the second column contained the number of cells that had been placed into each of the bins. These were then used to construct a histogram of counts versus cell size.

Results

Determination of a Constant Calibration Factor

Ideally, one would expect that a plot of cell size measured by flow cytometry versus cell size measured from photomicrographs would have a gradient of 1. However, as can be seen from Fig. 4.2 this is not the case. Figure 4.2 shows the data for the four types of unfixed cells measured in phosphate buffer (PBU cells). When the peak cell diameter determined by flow cytometry is plotted against the modal cell diameter measured from the photomicrographs, the gradient of the line $y = 0 + bx$, which is well fitted by linear least squares, is only 0.76. This suggests that calibration of the flow cytometric data with the latex calibration beads leads to a constant underestimation of the "true" size of these cells of 24%, presumably because cells

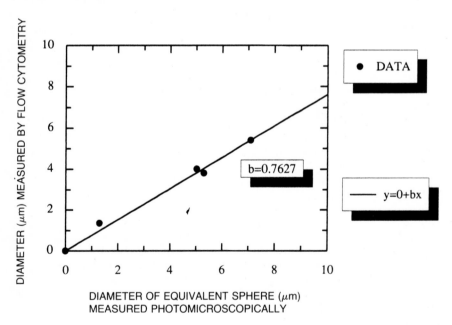

Fig. 4.2. Data for the unfixed cells in phosphate buffer (PBU) showing (in order of increasing size) *M. luteus* and yeast strains DCL2, DCL1 and BB1. The graph shows that calibration by latex beads gives an underestimation of true cell size. However, the gradient (*b*) of the line $y = 0 + bx$ gives an additional calibration factor that can be used to convert the apparent cell size measured by flow cytometry into a "true" cell size.

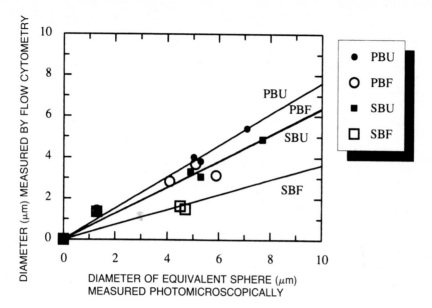

Fig. 4.3. Data from all four cell treatments for *M. luteus* and the three yeast strains. The data point for the yeast strain (BB1) is missing from the SBF plot because excessive clumping made interpretation of the flow cytometric data impossible. The gradients of the line ($y = 0 + bx$) for each treatment are as follows: PBU, 0.763; PBF, 0.637; SBU, 0.634; SBF, 0.365. It is obvious that a different calibration factor is required for each buffer and for fixed and unfixed cells in the same buffer.

do not scatter light as much as do latex beads of the same diameter. Since much of what is measured by flow cytometry is directly related to volume rather than to diameter (e.g. protein content), it should be noted that a 24% underestimation of diameter corresponds to a 56% underestimation of volume. Once the extent of the underestimation of cell size has been determined, however, it can be used as a constant factor for calibrating flow cytometric light scattering data obtained from microorganisms, apparently independently (within reason) of their size and nature.

When similar data for the other cell treatments are plotted (Fig. 4.3), it may be observed that the magnitude of the "constant factor" (ratio) is dependent on the buffer in which the cells are suspended and also on whether or not the cells have been fixed. The goodness of fit to a straight line is also clearly better for the unfixed cells. For each cell *treatment* used it will therefore be necessary to use a different calibration factor.

The effect of using the constant calibration factor for a *particular* treatment is a much improved measure of cell size. This can be seen in Fig. 4.4, which shows the photomicrographic measurements along with the flow cytometric data calibrated either with calibration beads alone or using both calibration beads and the relevant calibration factor. For the unfixed cells, the agreement of the latter flow cytometric measurements with the photomicrographic measurements is practically perfect (Fig. 4.4a,c). However, this is not the case for fixed cells (Fig. 4.4b,d).

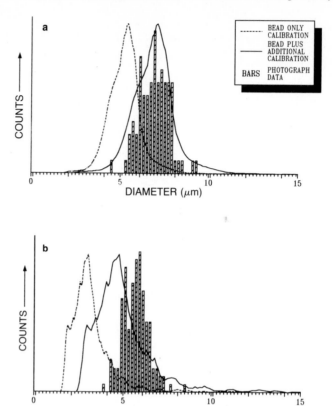

Fig. 4.4a–d. Effect of using the additional calibration factor obtained from the gradient for the relevant treatment shown in Fig. 4.3. By using the additional calibration factor a more accurate measure of cell size is obtained. **a** PBU cells; **b** PBF cells; **c** SBU cells; **d** SBF cells. It should be noted that even when the additional calibration factor is used the overlap of the data for the fixed cells is not perfect. The data shown in **a–c** are for yeast strain BB1, whilst **d** shows data for the DCL2 yeast. In each graph the flow cytometric data represent size measurements made on approximately 10 000 cells while the data from the photomicrographs represent only 100–200 cells. In order to plot these together the absolute counts have been deleted from the ordinate axes.

Effect of Fixing the Cells

When the cells were fixed with 70% ethanol some clumping occurred. This hindered cell size measurement from the photomicrographs to some extent as it was difficult to see the edges of cells in the clumps. More importantly though, it also affected the cell size distributions from the flow cytometer and, in the case of the largest strain of yeast considered (BB1), the clumping resulted in the peak channel number being outside the range encompassed by the calibration beads and so the data for the SBF cells of this yeast were lost. More, and larger, clumps were formed by the SBF cells (up to 28 cells per clump; average 30% of cells unclumped) than by the PBF cells (up to 9

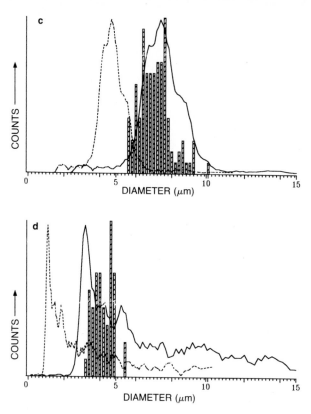

Fig. 4.4. *Continued*

cells per clump; average 61% of cells unclumped). The figures quoted here are for the three yeast strains only; with *M. luteus* no obvious clumping occurred with the PBF cells but the clumps formed by the SBF cells were so large that the number of cells per clump could not be determined. In Fig. 4.4d the effect of cell clumping on the flow cytometric data is clearly visible: a "tail" to the cell distribution is produced to the right of the unclumped cells in the main peak.

Another effect of fixing the yeast cells was to reduce their diameters. This was apparent both from the flow cytometry data and also from the photomicrographs, although the effect on the flow cytometric data was greater. No such reduction in cell size was observed in the case of *M. luteus*. The average reduction in the apparent cell diameter on fixing for the PBU cells of the three yeast strains was 13% when measured from the photomicrographs but was 25% when measured by flow cytometry. For the SBU cells the reduction was 16% when measured from the photomicrographs but 54% when measured by flow cytometry.

Related to the reduction in cell size caused by fixing with ethanol was a third effect on the yeast cells, namely the effect on the axial ratios of the cells. As can be seen in Fig. 4.5 the axial ratio of the cells increased after

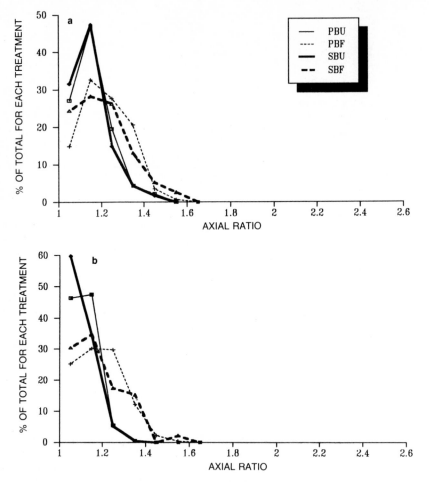

Fig. 4.5a–d. The effect of fixing the cells with 70% ethanol on their axial ratios. **a** BB1 yeast; **b** DCL2 yeast; **c** DCL1 yeast; **d** *M. luteus*.

fixing. This indicates that although the cells do shrink in both length and width (data not shown), they do not do so in equal proportions.

One final effect that was noticed following fixing the cells in suspension buffer was that in some cases the contents of the cells were released. This can clearly be seen (along with the other effects of fixing) in Fig. 4.6.

Finally, Fig. 4.7 shows the relationship between forward angle light scattering (size) and FITC staining (for protein content) for fixed and unfixed samples of *M. luteus*. Steen (1990) states that light scattering is proportional to the total protein content of the cells, and although there is some scatter on the plots, this seems generally to be true for both unfixed and fixed cells of *M. luteus*.

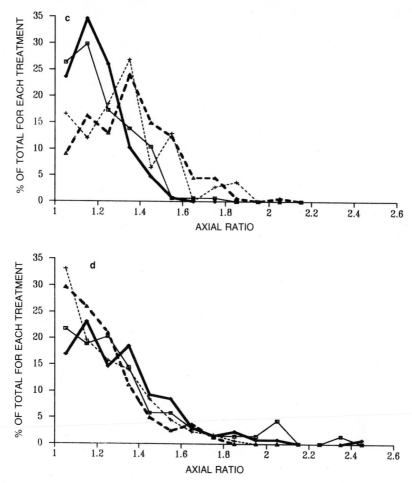

Fig. 4.5. *Continued*

Discussion

For a variety of flow cytometric applications it is essential to express cell sizes in absolute rather than relative terms. Figures 4.2–4.4 show that the amount of light scattered by a cell is always less than that of a latex bead of the same diameter, but by a constant fraction. The linearity of the data in Figs. 4.2 and 4.3 shows that the error is a constant *fraction* and not a constant *amount*. By comparing cell size distributions measured by flow cytometry with those obtained from photomicrographs we have obtained a set of additional calibration factors that enable one to convert bead-calibrated flow cytometric data to "true" cell sizes. Thus we have the convenience of a simple bead calibration coupled with an improved determination of cell size.

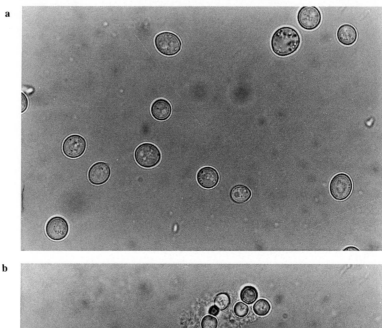

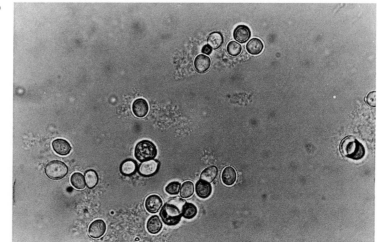

Fig. 4.6a,b. Photomicrographs of the BB1 yeast. **a** Unfixed yeast in suspension buffer (SBU yeast); **b** fixed yeast in suspension buffer (SBF yeast). Following the fixation step of the cell preparation several differences can be seen: the cells are smaller, large clumps of cells are formed and some cell contents have escaped into the medium.

It is apparent, however (Fig. 4.3), that a different calibration factor is required for each cell treatment (sample preparation) used.

The fact that fixing cells with 70% ethanol decreases their size as determined both by flow cytometry and by photomicrograph measurement means that one cannot determine "true" cell size distributions once cells have been fixed. This means, for example, that one cannot directly relate measurements of protein or DNA content obtained from flow cytometry of fixed cells to the cellular volume.

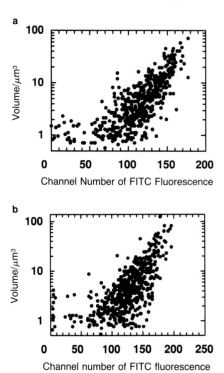

Fig. 4.7a,b. Relationship between volume and protein content of *M. luteus*. **a** Unfixed cells; **b** ethanol-fixed cells.

As shown in Fig. 4.4 the calibration factor improves the accuracy of cell size determination for all samples. However, while the cell size distributions for the unfixed (PBU and SBU) cells measured from the photomicrographs coincided almost exactly with the relevant flow cytometric data obtained by the improved calibration method, there was still a substantial discrepancy between the cell size distributions obtained by the two different methods for the PBF and SBF cells.

Flow cytometry has the prerequisite that the cell suspension of interest should consist of single cells, and, as shown here, the extent to which this is true can be strongly affected by fixing. The degree to which the cells are affected by fixing depends to some extent on which pre-fixation buffer is used. The correct interpretation of flow cytometric analyses therefore requires the determination of the effects of all stages of sample preparation on the cells.

We conclude that only in the case of unfixed cells is it possible to obtain a reliable value for their diameter by measuring the extent to which they scatter light at low angles in a flow cytometer.

Acknowledgement

We are indebted to the SERC for financial support of this work.

References

Alanen KA, Klemi PJ, Joensuu H, Kujari H, Pekkala E (1989) Comparison of fresh, ethanol-preserved, and paraffin-embedded samples in DNA flow cytometry. Cytometry 10:81–85

Boye E, Løbner-Olesen A (1990) Flow cytometry: illuminating microbiology. New Biol 2: 119–125

Boye E, Løbner-Olesen A (1991) Bacterial growth control studied by flow cytometry. Res Microbiol 142:131–135

Boye E, Steen HB, Skarstad K (1983) Flow cytometry of bacteria: a promising tool in experimental and clinical microbiology. J Gen Microbiol 129:973–980

Davey CL, Kell DB, Dixon NM (1990a) SKATFIT: A program for determining the mode of growth of individual microbial cells from flow cytometric data. Binary 2:127–132

Davey CL, Dixon NM, Kell DB (1990b) FLOWTOVP: A spreadsheet method for linearising flow cytometric light-scattering data used in cell sizing. Binary 2:119–125

Dean PN (1990) Data processing. In: Melamed MR, Lindmo T, Mendelsohn ML (eds) Flow cytometry and sorting. Wiley-Liss, New York, pp 415–444

Frelat G, Laplace-Builhe C, Grunwald D (1989) Microbial analysis by flow cytometry: present and future, In: Yen A (ed) Flow cytometry: advanced research and clinical applications, vol II, CRC Press, Boca Raton, FL, pp 255–279

Grant EH, Sheppard RJ, South GP (1978) Dielectric behaviour of biological molecules in solution. Clarendon Press, Oxford

Horan PK, Muirhead KA, Slezak SE (1990) Standards and controls in flow cytometry. In: Melamed MR, Lindmo T, Mendelsohn ML (eds) Flow cytometry and sorting. Wiley-Liss, New York, pp 397–414

Kachel V, Fellner-Feldegg H, Menke E (1990) Hydrodynamic properties of flow cytometry instruments. In: Melamed MR, Lindmo T, Mendelsohn ML (eds) Flow cytometry and sorting. Wiley-Liss, New York, pp 27–44

Kell DB, Davey HM, Kaprelyants AS, Westerhoff HV (1991) Quantifying heterogeneity: flow cytometry of bacterial cultures. Ant van Leeuwenhoek 60:145–158

Kogoma T, Skarstad K, Boye E, von Meyenburg K, Steen HB (1985) RecA protein acts at the initiation of stable DNA replication in *rnh* mutants of *Escherichia coli* K-12. J Bacteriol 163:439–444

McCoy JP, Lovett EJ (1989) Basic principles in clinical flow cytometry. In: Keren DF (ed) Flow cytometry in clinical diagnosis, ASCP Press, Chicago, pp 12–40

Muirhead KA, Horan PK, Poste G (1985) Flow cytometry: present and future. Biotechnology 3:337–356

Mullaney PF, Dean PN (1970) The small angle light scattering of biological cells. Biophys J 10:764–772

Murphy RF, Chused TM (1984) A proposal for a flow cytometric data file standard. Cytometry 5:553–555

Paau AS, Cowles JR, Oro J (1977) Flow-microfluorometric analysis of *Escherichia coli*, *Rhizobium meliloti* and *Rhizobium japonicum* at different stages in the growth cycle. Can J Microbiol 23:1165–1169

Salzman GC, Singham SB, Johnston RG, Borhen CF (1990) Light scattering and cytometry. In: Melamed MR, Lindmo T, Mendelsohn ML (eds) Flow cytometry and sorting. Wiley-Liss, New York, pp 81–107

Scheper Th, Hoffmann H, Schugerl K (1987) Flow cytometric studies during culture of *Saccharomyces cerevisiae*. Enzyme Microb Technol 9:399–405

Shapiro HM (1983) Multistation multiparameter flow cytometry: A critical Review and rationale. Cytometry 3:227–243

Shapiro HM (1988) Practical flow cytometry. Alan R Liss, New York

Shapiro HM (1990) Flow cytometry in laboratory microbiology: New directions. ASM News 56:584–588

Skarstad K, Steen HB, Boye E (1985) *Escherichia coli* DNA distributions measured by flow cytometry and compared with theoretical computer simulations. J Bacteriol 163:661–668

Skarstad K, Boye E, Steen HB (1986) Timing of chromosome replication in individual *Escherichia coli* cells. EMBO J 5:1711–1717

Steen HB (1990) Flow cytometric studies of microorganisms. In: Melamed MR, Lindmo T, Mendelsohn ML (eds) Flow cytometry and sorting. Wiley-Liss, New York, pp 605–622

Steen HB, Lindmo T (1985) Differential of light scattering in an arc-lamp-based epi-
   illumination flow cytometer. Cytometry 6:281–285
Steen HB, Lindmo T, Stokke T (1989) Differential light-scattering detection in an arc lamp-
   based flow cytometer. In: Yen A (ed) Flow cytometry: advanced research and clinical
   applications, vol I. CRC Press, Boca Raton, FL, pp 63–80
Steen HB, Skarstad K, Boye E (1990) DNA measurements of bacteria. In: Darzynkiewicz Z,
   Crissman HA (eds) Flow cytometry, Academic Press, London, pp 519–526
Wittrup KD, Mann MB, Fenton DM, Tsai LB, Bailey JE (1988) Single-cell light scatter as a
   probe of refractile body formation in recombinant *Escherichia coli*. Biotechnology 6:423–426

Chapter 5

Uses of Membrane Potential Sensitive Dyes with Bacteria

David Mason, Richard Allman and David Lloyd

Introduction

Dyes as Indicators of Membrane Potential

An important property of all biological membranes is that they are selectively permeable to a variety of cations and anions (including the principal cellular ions H^+, Na^+, K^+ and Cl^-), so that the different ions tend to move down their concentration gradients through the membrane at different rates. These two characteristics, selective permeability and ionic concentration gradients, lead to a difference in electric potential between the inside and the outside of a cell.

The magnitude of the resultant membrane potential across a cell surface is given by the Goldman equation, in which the concentrations of the ions are weighted in proportion to their permeability constants:

$$E = \frac{RT}{F} \ln \frac{P_{K^+}[K^+]_o + P_{Na^+}[Na^+]_o + P_{Cl^-}[Cl^-]_i}{P_{K^+}[K^+]_i + P_{Na^+}[Na^+]_i + P_{Cl^-}[Cl^-]_o}$$

$$= 59 \log_{10} \frac{P_{K^+}[K+]_o + P_{Na^+}[Na^+]_o + P_{Cl^-}[Cl^-]_i}{P_{K^+}[K^+]_i + P_{Na^+}[Na^+]_i + P_{Cl^-}[Cl^-]_o}$$

The "o" and "i" denote the concentrations outside and inside the cell respectively and, by convention, the membrane potential is expressed as inside relative to outside. Due to their opposite charges $[K^+]_o$, $[Na^+]_o$ and $[Cl^-]_i$ are placed in the numerator; conversely $[K^+]_i$, $[Na^+]_i$ and $[Cl^-]_o$ are in the denominator. This equation adequately describes the potential at any point where the concentrations and permeabilities are known. If the concentrations of the ions are equilibrium concentrations, then the potential will be the equilibrium potential, which will be retained unless outside energy is applied to the cell, or the concentration of an ion is altered. If the concentrations are not equilibrium ones, then the equation will describe the present potential but not the potential at any future time.

The significance of•these electrical membrane potentials has long been realized by life scientists for their involvement and importance in a number of cellular functions such as energy production, movement, signalling and development.

A direct method of measurement of the membrane potential comes from the use of microelectrodes. The membrane potential is defined as that between two reference electrodes immersed in the bulk solutions in contact with the two sides of the membrane. However, the small scale of bacteria, organelles and vesicles makes the application of this method of measurement limited. Hence most techniques for investigating these systems are still indirect ones.

One of the most popular of the indirect methods for measuring the membrane potential in bacteria and small vesicles is by measurement of the distribution of permeant cations such as K^+, Rb^+ or the synthetic lipophilic species tetraphenylphosphonium (TPP^+), dibenzyl dimethyl ammonium (DDA^+) or triphenylmethylphosphonium ($TPMP^+$). These are either used as radiolabelled tracers, e.g. $^{86}Rb^+$ (Schuldiner and Kaback 1975), or must be sensed in the medium by ion-selective electrodes, e.g. TPP^+ (Kamo et al. 1979; Lolkema et al. 1982; Eriok and Webster 1990). Permeant cations provide a method for measuring the membrane potential when the interior of the cells is negative. If the interior is positive as in alkalophilic bacteria and in everted membrane vesicles, permeant anions must be used. The technique is based on a fundamental thermodynamic principle that for a vesicular system at equilibrium, the electrochemical gradient of permeable ions vanishes, and the membrane potential can then be calculated from the equilibrium concentrations of the ions using the Nernst equation:

$$\Delta \Psi = -2.3(RT/ZF) \log (a_{in}/a_{out})$$

Bacterial membranes have been found to be virtually impermeable to K^+; however with the addition of an ionophore specific for K^+. (e.g. valinomycin) the membrane can be permeabilized to K^+. These ions, or their analogues (e.g. $^{86}Rb^+$), will redistribute themselves across the membrane of valinomycin-treated bacterial cells. At equilibrium the ratio of their concentrations across the membrane equals the membrane potential. The membrane potential is thus calculated from the distribution of K^+ or $^{86}Rb^+$ using the Nernst equation (Zilberstein et al. 1979; Bakker 1982). This method has been used for *Escherichia coli* (Booth et al. 1979; Ahmed and Booth 1981).

The main disadvantage of this method is that the K^+ concentration in the medium must be kept low (less than 50 $\mu mol/l$). This reduces the size of the K^+ current at steady state, which affects the measured potential. This can be too restrictive for some experiments. Another drawback to this technique is that it requires an accurate determination of the cell or vesicle volume.

Use of synthetic lipophilic ions, particularly $TPMP^+$, DDA^+ and TPP^+, to measure membrane potential has been popular with a number of investigators. DDA^+ requires the presence of a lipophilic anion, tetraphenylboron (TPB^-) (Muratsuga et al. 1977; Shinbo et al. 1978; Kamo et al. 1979) to accelerate its otherwise slow diffusion across membranes. As a result TPP^+, which diffuses far more rapidly, is favoured. The advantage of this approach is that it can be used with growing cells. However, in some cases these ions have been found to be inhibitory to membrane-bound enzymes and transport systems, and this limits their use. Use of the ions at low concentrations (less than $3\,\mu mol/l$) can go some way towards minimizing this problem. Nevertheless a significant proportion of these ions is inevitably bound to the membrane. This complicates matters when calculating the internal concentration from the measured uptake of the ion, as it is important to correct for the bound ion concentration. Excess binding to the membrane and other components is a common problem with all lipophilic cations, and hence calculated values of membrane potential based on the distribution of these ions are often overestimates.

The use of optical membrane potential probes has become popular in recent years. These probes can be divided into two categories: intrinsic and extrinsic. Intrinsic probes are naturally occurring chromophores (e.g. carotenoid pigments) that respond directly to membrane potential. Extrinsic probes are added to the membrane preparation, the advantage being that they can be used for many different types of membrane.

Of primary importance when choosing an extrinsic probe is that it is nontoxic to the type of cells with which it is to be used, and does not interfere with normal membrane function. To facilitate this the probe should have both a high extinction coefficient and a high quantum yield, enabling the use of very low quantities of the probe. Additional properties to consider are that it should not interact with other reagents used in the experiment, and that to avoid optical interference the absorption and emission characteristics of the probe should be well separated from those of any natural pigments occurring in the membrane.

Cyanine dyes have proved to be one of the most widely used group of extrinsic probes. These dyes are lipophilic with a delocalized cationic charge:

The shorthand nomenclature DiYCn $(2m + 1)$ was devised by Sims et al. (1974), where Y is the isopropyl member of the ring structure, which can be oxygen (O) or sulphur (S). The lengths of the alkyl side chains (n) affect the lipid solubility, and the number of methane groups (m) affects the fluorescence spectral characteristics. The fluorescence intensity change is proportional to the membrane potential calculated from the constant field equation in both human and salamander (*Aphiuma*) cells (Hoffman and Laris 1974). In the case of *Aphiuma* cells the calculated membrane potential also agreed with values obtained directly using microelectrodes. These dyes have been used extensively to determine membrane potential in a variety of cells and vesicles including red blood cells (Sims et al. 1974), mouse ascite tumour cells (Philo and Eddy 1978; Eddy 1989), cultured mammalian (Schwann) cells (Hargittai et al. 1991), yeast cells (Pena et al. 1984), mitochondria (Petit et al. 1990), and some bacteria (Zaritsky et al. 1981).

Another popular optical probe is the laser dye rhodamine 123, which is also lipophilic with a delocalized cationic charge:

Rhodamine 123 has most commonly been used to determine membrane potential in mitochondria (Ronot et al. 1986; Skowronek et al. 1990). However it has also been used for a variety of other cells including bacteria and murine leukaemia cells (Kessel et al. 1991).

The use of these probes gives the advantages of both high sensitivity to small changes in membrane potential, and rapid response. However the user is also faced with two major drawbacks, resulting from the fact that the mechanism by which the probes respond to membrane potential is largely unknown. Firstly, membrane potential is not the only factor that governs the response of these probes; other factors, including the effects of solvents, pH quenchers and self-quenching, should be considered. Thus there is often uncertainty in quantitative estimation of membrane potential using optical probes because of multifactorial effects. Secondly, it is not possible to obtain a value for the membrane potential directly using these probes as the mechanism of response is unknown. One solution is to calibrate the fluorescence changes of the probe against known potassium diffusion potentials.

In the work described here we set out to calibrate the variations in optical probe uptake (and consequent fluorescence intensity) in bacteria as a function of membrane potential. We quantitated the uptake of two probes, a cyanine dye 3,3-dihexyloxocarbocyanine iodide [DioC$_6$(3)], and rhodamine 123 (rh123), by gram-negative and gram-positive bacteria under controlled conditions whereby the membrane potential voltages were fixed at specific values.

Membrane Potential Sensitive Dyes as Indicators of Antibiotic Sensitivities

Fluorescent membrane potential probes may also be used to study the effects of drugs on bacteria.

Within the last 20 years bacterial antibiotic resistance has become a major problem and the economic costs are enormous (Lambert 1988). Multiply resistant strains of *Staphylococcus aureus* (MRSA) and *Pseudomonas aeruginosa* are a constant problem in hospitals. Of particular concern are the methicillin-resistant MRSA against which few agents are available; strains have recently developed which are also resistant to ciprofloxacin (Shalit et al. 1989). Emerging pneumococcal strains with resistance to a number of antibiotics (penicillin, tetracycline, chloramphenicol and trimethoprim-sulphamethoxazole) represent an enormous challenge to existing technologies (Pallares et al. 1987; Klugman and Koornhof 1988). Antibiotics such as streptomycin, nalidixic acid, fusidic acid and rifampicin are renowned for promoting rapid emergence of resistance.

Control of the development of resistance relies on usage of an antibiotic to a strict protocol (Bendall et al. 1986). In order to establish and maintain a usage protocol, determination of susceptibility of the prevalent bacteria to antibiotics is crucial. The most commonly used methods for this are the disc diffusion test (Stokes and Ridgeway 1987; Piddock 1990) and the tube titration method (Stokes and Ridgeway 1987). These methods are effective, but they are not without their problems:

1. Standardization of the tests between clinical laboratories is difficult; in the case of the tube titration method the inoculum size may differ from one test to another and this can significantly affect results, especially with β-lactam antibiotics.
2. Interpretation of results from the two types of test can vary.
3. The time involved (often a matter of days between the start of susceptibility testing and obtaining the results) may prejudice survival.

Measurement of the uptake of fluorescent probes has been applied to the detection of drug resistance/sensitivity in mouse thymocytes and murine leukaemia cells (Kessel et al. 1991; Oyama et al. 1991). It has recently been suggested that measurement of the uptake of rh123 may be useful in evaluating the efficacy of antitumour drugs on carcinoma cells (Shinomiya et al. 1992).

We have started looking at how the addition of an antibiotic to bacterial cells may affect the uptake of rh123 and consequential fluorescence. The example we use is the treatment of *E. coli* cells with the aminoglycoside, kanamycin.

Methods

Membrane Potential and Dye Uptake Measurements

The bacterial strains chosen were *Escherichia coli* C600 and *Staphylococcus aureus* NCTC 6571. In all experiments with *E. coli* final incubation mixtures included 1 mmol/l EDTA to enhance permeability. Cultures of these strains were grown in medium at 37 °C to the mid-exponential phase of growth before any measurements were carried out. Measurement of the membrane potential was based on the potassium diffusion technique. The external potassium concentration was fixed at different values by suspension of the cells in a number of buffered standards, each containing a different concentration of KCl within the range 10–200 mmol/l. Cells were removed from these standards by centrifuging through a phthalate oil mixture (consisting of 1 part dinonylphthalate to 2 parts dibutylphthalate (v/v), which gave a density of 1.02 g/ml). The pellet, supernatant and oil layers were separated, and the cells in the pellet sonicated for 30 s and resuspended in distilled water. The K^+ content of the supernatant and of disrupted cells was measured using a flame photometer. The volume of the cells from both strains was determined using a Coulter counter fitted with a 30 μm orifice.

Fluorescence measurements were carried out on a spectrophotofluorimeter (Applied Photophysics, UK) and an Argus Flow Cytometer, fitted with an FITC filter block (Skatron, Norway). Cells of either strain were suspended in the various potassium standards and stained with rh123 (0.05 μg/ml) or DioC$_6$(3) (0.1 μg/ml). For these experiments we chose to study two strains of *E. coli* sensitive (C600) or resistant (C600 Pkt 231) to kanamycin.

Each strain was grown in broth culture supplemented with glucose at 37 °C to the mid-exponential phase of growth. At this time a sample was removed from the culture and antibiotic was added to the remaining part of the culture to give a final concentration of 25 μg/ml. The culture was then incubated for a further 4 min before another sample was removed.

Bacteria were harvested by centrifugation for 1 min in an Eppendorf microcentrifuge. The resulting pellet was resuspended in a buffered solution (pH 7.6), containing 3 mmol/l glucose and 1 mmol/l EDTA and prewarmed to 37 °C. Rh123 was then added to give a final concentration of 0.25 μg/ml. The samples were incubated for a further 20 min before analysis.

Results

Measurements of membrane potential using values of intracellular and extracellular K$^+$ concentrations determined by flame photometry gave results in reasonable agreement with theoretical predictions (Fig. 5.1). Thus, according to the Nernst equation, a 10-fold increase in external K$^+$ concentration should produce a 60 mV change in membrane potential at room

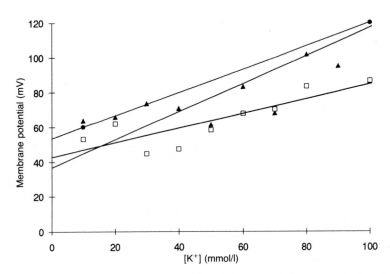

Fig. 5.1. Comparison of membrane potentials of *S. aureus* (*triangles*) and *E. coli* C600 (*squares*). Typical results as determined using flame photometry. According to the Nernst equation, at 25 °C a 10-fold increase in extracellular [K$^+$] should produce a change in membrane potential of 60 mV. *Circles*, theoretical predictions.

a

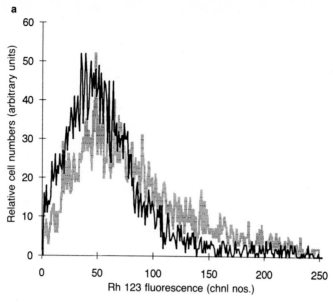

b

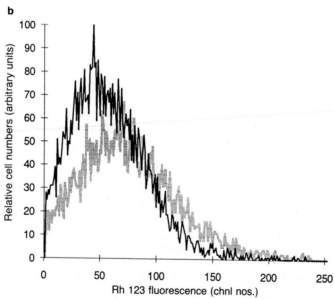

Fig. 5.2a,b. Effect of 15 μmol/l CCCP (*m*-Cl carbonyl cyanide phenylhydrazone) on the fluorescence of rh123-stained organisms: **a** *S. aureus*; **b** *E. coli*. *Grey lines*, before treatment; *black lines*, after treatment. Washed organisms were incubated with 30 mmol/l glucose and 0.25 μg/ml rh123 for 30 min before adding the uncoupler. Flow cytometric determination of fluorescence was then performed immediately.

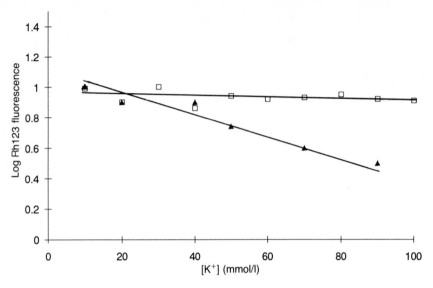

Fig. 5.3. Fluorimetric measurements of dye uptake by *S. aureus* (*triangles*) and *E. coli* C600 (*squares*) stained with rh123 (0.05 μg/ml). Excitation was 505 nm, emission was 520 nm.

temperature. The values obtained experimentally were 49 mV and 45 mV for *S. aureus* and *E. coli* respectively.

When using the flow cytometer the fluorescence measured was that from individual bacteria. So when organisms became hyperpolarized an increase in fluorescence was observed. If the cells became depolarized then fluorescence decreased. Figure 5.2 shows the effects of an uncoupler of energy conservation on rh123 fluorescence in *S. aureus* and *E. coli*.

Fluorescence measurements of cell suspensions in cuvettes using the fluorescence spectrometer actually measure the extracellular component (i.e. the fluorescence from the unbound dye in solution). Hence when cells become hyperpolarized increased dye uptake results in decreased fluorescence. Conversely when cells become depolarized, dye molecules released into free solution give increased fluorescence. Measurements of fluorescence changes by this method are inherently less sensitive than those using flow cytometry as small variations in an intense fluorescence background may be significant. This is shown in Fig. 5.3; only for *S. aureus* with rh123 were we able to demonstrate a clear correlation between increased $[K^+]_o$ and decreased extracellular fluorescence.

Figures 5.4 and 5.5 show the results of similar experiments using flow cytofluorimetric measurements. As $[K^+]_o$ was increased, both *S. aureus* and *E. coli* exhibited increased fluorescence when incubated with rh123. For *E. coli* the use of $DioC_6(3)$ confirmed this relationship, but for *S. aureus* increasing $[K^+]_o$ gave decreased fluorescence intensities.

The results shown in Fig. 5.6 indicate that on addition of antibiotic the kanamycin-sensitive *E. coli* C600 cells exhibited a larger decrease in fluorescence than the resistant C600 Pkt 231 cells.

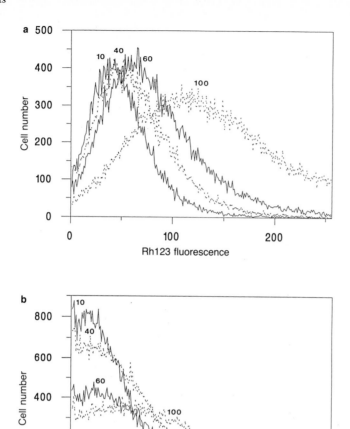

Fig. 5.4a–d. Typical fluorescence histograms obtained using flow cytometry. **a** *S. aureus* stained with rh123 (0.05 μg/ml); **b** *E. coli* C600 stained with rh123 (0.05 μg/ml). (*Continued overleaf.*)

A similar experiment was conducted with two strains of *S. aureus*: one ampicillin-resistant (NCTC 9682) and the other ampicillin-sensitive (NCTC 6571). These results (Table 5.1) were similar to those for *E. coli* in that the sensitive strain exhibited a greater reduction in fluorescence than the resistant strain.

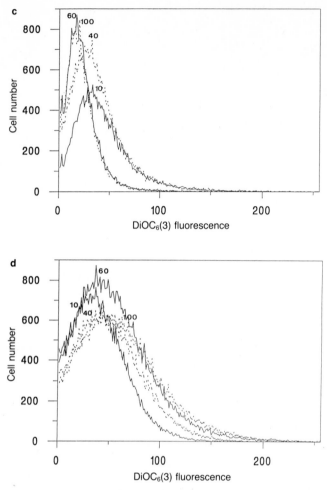

Fig. 5.4. *Continued.* **c** *S. aureus* stained with DioC$_6$(3) (0.1 µg/ml); **d** *E. coli* C600 stained with DioC$_6$(3) (0.1 µg/ml). In each case histograms are shown for organisms suspended in 10, 40, 60, 100 mmol/l KCl solutions.

Discussion

Discrepancies between the experimentally determined and theoretical membrane potential values are as follows: (1) there may have been some leakage of intracellular K$^+$ during processing; (2) during centrifugation procedures some organisms may have escaped recovery; (3) the calculation ignores the possibility of other ion movements (especially as the Goldman equation highlights the importance of Cl$^-$ which therefore really should be considered also).

The fluorescence results indicate that rh123 is a very suitable probe for the determination of membrane potential (with approximately one channel in the fluorescence histograms corresponding to 1 mV) in *S. aureus*, and also

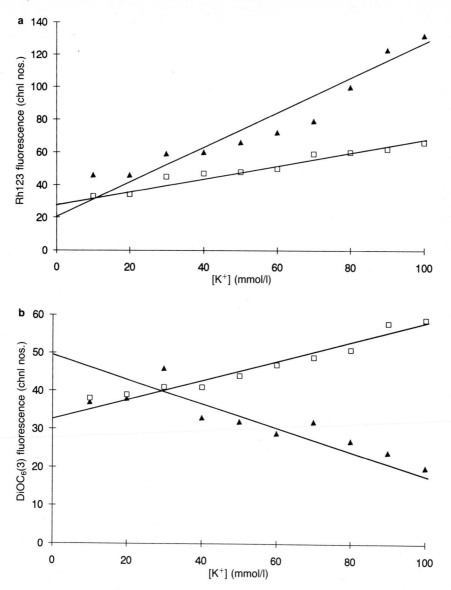

Fig. 5.5. Plots of the median channel numbers of the flow cytometric histograms versus external $[K^+]$ for: **a** *S. aureus* (*triangles*) and *E. coli* C600 (*squares*) stained with rh123 (0.5 μg/ml); **b** *S. aureus* (*triangles*) and *E. coli* C600 (*squares*) stained with $DioC_6(3)$ (0.1 μg/ml).

in *E. coli*, provided that EDTA is present. Differences in dye permeability are probably due to the properties of the outer membrane of *E. coli*.

Fluorescence changes observed with $DioC_6(3)$ for the various external K^+ concentrations were much lower than those with rh123. There was also some evidence of toxicity of $DioC_6(3)$ to *S. aureus*. This was revealed most clearly in the flow cytometric data. In toxicity experiments the minimum inhibitory

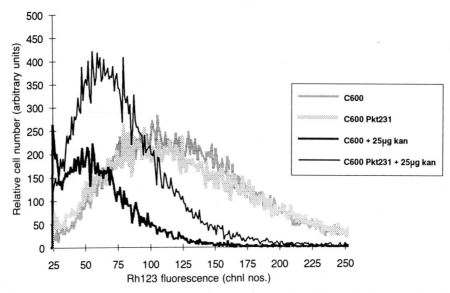

Fig. 5.6. Effects of kanamycin (kan) on the fluorescence of sensitive and resistant strains of rh123-stained *E. coli*. The procedure was as in Fig. 5.2, except that 25 µg/ml antibiotic was added to growing cultures before washing and staining. C600 Pkt231 was the resistant strain.

Table 5.1. FL1 modal and median channel numbers for ampicillin-treated *S. aureus*

Sample	*S. aureus* NCTC 6571 (ampicillin-sensitive)		*S. aureus* NCTC9682 (ampicillin-resistant)	
	Modal	Median	Modal	Median
Untreated cells	58	66	48	57
Antibiotic-treated cells	48	61	47	54

concentration of DioC$_6$(3) for actively growing cultures of *S. aureus* was 1 µg/ml, whereas *E. coli* were able to tolerate 5 mg/ml. The mechanism of the toxicity to *S. aureus* is not well understood; toxicity experiments were conducted in the dark to exclude the possibility of photodynamic damage. It seems possible that the presence of an outer membrane in *E. coli* affords protection against a similar fate. It has been reported that cyanine dyes have acted as uncouplers of energy conservation in some eukaryotic systems (Chused et al. 1986), although the concentrations used were much higher than those used here. The observed depolarization of the *S. aureus* membranes suggests proton conduction as a possible mechanism.

We conclude that of the two dyes used, rh123 exhibited a greater response to membrane potential and is therefore the more useful. However DioC$_6$(3) is only one of a large family of similar compounds, and it seems possible that another dye chosen from this range might be more satisfactory. Our results clearly show that permeability towards specific dyes is likely to differ between bacterial species, so that no universal ground rules can be established.

Histograms of the fluorescence distributions for both dyes, and for both bacterial species, were much more dispersed than cell size distributions (as determined using forward light scatter measurements). This discrepancy may arise from dyes binding to cellular targets other than the plasma membrane; broad fluorescence distributions will then reflect differences in numbers of binding sites between individuals of the population. These non-specific target sites may be analogous to "lipid-rich domains" in eukaryotic cells.

In the experiments described, the membrane potential was assumed to be the sole determinant of the response of the probe, which was "fixed" at a number of values by varying the external $[K^+]$. However if this were exactly so, fluorescence frequency distributions would be less dispersed.

In summary, although it is possible to assign approximate membrane potential values to rh123 fluorescence intensity measurements, interpretation of the dispersion of frequency distributions is not possible without further work.

Although there are detectable differences in antibiotic-induced fluorescence changes between sensitive and resistant strains, this difference is not always very apparent.

The aim of this work is to develop a technique for the rapid detection of antibiotic resistance/sensitivity in bacteria by flow cytometry. In "blind" experiments (results not shown) using kanamycin with the two strains of *E. coli* successful identification of the resistant and sensitive strains was shown to be possible. However work is still required to develop this method further so that more pronounced changes in fluorescence are specifically produced by the effects of antibiotics on the membrane potential of sensitive strains. The present method allows 20 min for binding of rh123. Depending on the bacterial species employed, maximum binding is attained within 5–30 min (Ronot et al. 1986). Therefore it is possible that to obtain larger differences in fluorescence, incubation time of the samples should be increased. Increasing the concentrations of rh123 is another option; concentrations of up to $5 \mu g/ml$ ($13 \mu mol/l$) have been used previously (Matsuyama 1984). However it is desirable to keep the concentration as low as possible ($<1 \mu mol/l$) to avoid energy-independent dye accumulation and self-quenching (Kaprelyants and Kell 1992).

Acknowledgements

This work was supported by grants from the Ministry of Defence and the Welsh Office.

References

Ahmed S, Booth IR (1981) Quantitative measurements of the proton-motive force and its relation to steady state lactose accumulation in *Escherichia coli*. Biochem J 200:573–581

Bakker EP (1982) Membrane potential in a potassium transport-negative mutant of *Escherichia coli* K-12. Biochim Biophys Acta 681:474–483

Bendall MJ, Ebrahim S, Finch RG, Slack RCB, Towner KJ (1986) The effect of an antibiotic policy on bacterial resistance in patients in geriatric wards. J Med 60:849–854

Booth JR, Mitchell WJ, Hamilton WA (1979) Quantitative analysis of proton-linked transport systems. Biochem J 182:687–696

Eddy A (1989) Use of carbocyanine dyes to assay membrane potential of mouse ascites tumour cells. Meth Enzymol 172:95–101

Eriok BJS, Webster DA (1990) Respiratory-driven Na^+ electrical potential in the bacterium *Vitreoscilla*. Biochemistry 29:4734–4739

Hargittai PT, Youmans SJ, Lieberman EM (1991) Determination of the membrane potential of cultured mammalian Schwann cells and its sensitivity to potassium using a thiocarbocyanine fluorescent dye. Glia 4:611–616

Kamo N, Muratsuga M, Hongoh R, Kobatake Y (1979) Membrane potential of mitochondria measured with an electrode sensitive to tetraphenyl phosphonium and relationship between proton electrochemical potential and phosphorylation potential in steady state. J Membr Biol 49:105–121

Kaprelyants AS, Kell DB (1992) Rapid assessment of bacterial viability and vitality using Rhodamine 123 and flow cytometry. J Appl Bacteriol 72:410–422

Kessel D, Beck WT, Kukuruga D, Schulz V (1991) Characterization of multidrug resistance by fluorescent dyes. Cancer Res 51:4665–4670

Klugman KP, Koornhof HJ (1988) Bacteremic pneumonia caused by penicillin-resistant pneumococci. N Engl J Med 318:123–124

Lambert HP (1988) Clinical impact of drug resistance. J Hosp Inf 2 (Suppl A):135–141

Lolkema JS, Hellingwerf KJ, Konings WN (1982) The effect of "probe binding" on the quantitative determination of the proton-motive force in bacteria. Biochim Biophys Acta 681:85–94

Matsuyama T (1984) Staining of living bacteria with rhodamine 123. FEMS Microbiol Lett 21:153–157

Muratsuga M, Kamo N, Kurihara K, Kobatake Y (1977) Selective electrode for diebenzyl diemethyl ammonium cation as indicator of the membrane potential in biological systems. Biochim Biophys Acta 464:613–619

Oyama Y, Chikahisa L, Tomiyoshi F, Hayashi H (1991) Cytotoxic action of triphenyltin on mouse thymocytes: a flow-cytometric study using fluorescent dyes for membrane potential and intracellular Ca^{2+}. Jpn J Pharmacol 57:419–424

Pallares R, Gudiol F, Linares J et al. (1987) Risk factors and response to antibiotic therapy in adults with bacteremic pneumonia caused by penicillin-resistant pneumococci. N Engl J Med 317:18–22

Pena A, Uribe S, Pardo JP, Barbolla M (1984) The use of a cyanine dye in measuring membrane potential in yeast. Arch Biochem Biophys 231:217–225

Petit PX, O'Connor JE, Grunwald D, Brown SC (1990) Analysis of the membrane potential of rat and mouse liver mitochondria by flow cytometry and possible applications. Eur J Biochem 194:389–397

Philo R, Eddy AA (1978) The membrane potential of mouse ascites – tumour cells studied with the fluorescent probe 3,3-dipropyloxadicarbocyanine. Amplitude of the depolarization caused by amino acids. Biochem J 174:801–810

Piddock LJV (1990) Techniques used for the determination of antimicrobial resistance and sensitivity in bacteria. J Appl Bacteriol 68:307–318

Ronot X, Benel L, Adolphe M, Mounlou J (1986) Mitochondria analysis in living cells: the use of Rhodamine 123 and flow cytometry. Biol Cell 57:1–8

Schuldiner S, Kaback HR (1975) Membrane potential and active transport in membrane vesicles from *Escherichia coli*. Biochemistry 14:5451–5416

Shalit I, Berger SA, Gorea A, Frimerman H (1989) Widespread quinolone resistance among methicillin-resistant *Staphylococcus aureus*: isolates in a general hospital. Antimicrob Agents Chemother 33:593–594

Shinbo T, Kamo N, Kurihara K, Kobatake (1978) A PVC based electrode sensitive to DDA^+ as a device for monitoring the membrane potential in biological systems. Arch Biochem Biophys 187:414–419

Shinomiya N, Tsuru S, Katsura Y, Sekiguchi I, Suzuki M, Nomoto K (1992) Increased mitochondrial uptake of Rhodamine 123 by CDDP treatment. Exp Cell Res 198:159–163

Sims J, Waggoner AS, Wang C, Hoffman JF (1974) Studies of the mechanism by which cyanine dyes measure membrane potential in red blood cells and phosphatidylcholine vesicles. Biochemistry 13:3315–3329

Skowronek P, Krummeck G, Haferkump O, Rodel G (1990) Flow cytometry as a tool to discriminate respiratory competent and respiratory deficient yeast cells. Curr Genet 18: 265–267

Stokes EJ, Ridgeway GL (1987) Clinical microbiology, 6th edn. Edward Arnold, London

Zaritsky A, Kihara M, MacNab RM (1981) Measurement of membrane potential in *Bacillus subtilis*: a comparison of lipophilic cations, rubidium ion, and a cyanine dye as probes. J Membr Biol 63:215–231

Zilberstein D, Schudliner S, Padan E (1979) Proton electrochemical gradient in *Escherichia coli* cells and its relation to active transport of lactose. Biochemistry 18:669–673

Flow Cytometric Analysis, Using Rhodamine 123, of *Micrococcus luteus* at Low Growth Rate in Chemostat Culture

Hazel M. Davey, Arseny S. Kaprelyants and Douglas B. Kell

Introduction

In microbiology it is often necessary to determine the number of viable cells in a sample or culture of interest. This is usually achieved by plating out the sample (diluted as required) on to an agar plate (Postgate 1969; Hattori 1988). There are several problems associated with this technique, the greatest of which is the length of time required to obtain the results. For some slowly growing organisms (e.g. Mycobacteria) it may take in excess of a week to determine how many cells were "viable" in the original sample, and even when the sample contains fast-growing organisms and the plates are incubated under optimal growth conditions a minimum of overnight growth is usually required before the resulting colonies can be counted. For some clinical specimens even an overnight incubation may be too long to be of use and consequently many alternatives to plate counts have been proposed in order to decrease the time required to determine numbers of viable cells (Harris and Kell 1985).

Many of these so-called rapid methods involve the use of dyes to stain the cells. These include the DNA stain acridine orange (McFeters et al. 1991) that is used on the (rather doubtful) assumption that nucleic acids are degraded rapidly following cell death, such that "viable" cells are stained differently from "dead" ones (see Back and Kroll 1991). Other viability stains such as methylene blue exploit the ability of the intact membrane of viable cells to exclude the dye (Jones 1987; Stoicheva et al. 1989). Dye exclusion and DNA-staining methods rely mainly on microscopic examination of the stained cells to assess viability, and for this reason they are slow to perform and are prone to subjective error.

Plate counts, although usually considered to be a measure of viability, actually indicate only how many of the cells can replicate under the conditions provided for growth, which are likely to differ from those in the original sample (Roszak and Colwell 1987). Viability staining meanwhile provides information on how many of the cells can exclude the dye (i.e. how many of the cells have intact cell membranes). However, there is abundant evidence that cells may exist in a state intermediate between being unable to replicate on an agar plate, and being dead as judged by viability stains (Postgate 1976; Jones 1987; Roszak and Colwell 1987). With conventional viability staining, microscopic examination is used to determine whether the cells are alive. By this method a judgement of "alive" or "dead" is all that is possible.

An alternative method of determining the amount of dye taken up by each cell is the technique of flow cytometry (Shapiro 1988; Pollack and Ciancio 1990). As described by many other authors in this volume, and recently reviewed elsewhere (Kell et al. 1991), flow cytometry is a technique that enables measurements to be made very rapidly on a large number of cells. Since the measurements are made on individual cells rather than on populations, differences in dye uptake between cells can be assessed quantitatively. Using supra-vital stains, estimates of the "degree of viability" of individual cells can thus be made, allowing one to *quantify* the hetero-geneity of the cell population (Kell et al. 1991).

In a flow cytometer particles pass rapidly in single file through an illumination zone, and appropriate detectors measure the amount of light scattered or, via suitable filters, the fluorescence of the particle (Shapiro 1988). The data are converted into pulses, the magnitude of which represent the amount of scattered light, and/or the fluorescence of each cell as it passes through the illumination zone. The pulses are "binned" into channels that increase in number with increasing levels of scattered light or fluores-cence. The data are then usually plotted as a histogram on which the abscissa represents channel numbers and the ordinate represents the number of cells measured in each channel (Dean 1990).

In conventional flow cytometers the source of the illumination is a laser (Shapiro 1988). However, given that both the cell volume and the DNA content of bacteria are some 1000-fold less than those of higher eukaryotic cells, laser-based flow cytometers have until recently proved unsuitable for the study of microorganisms (Steen et al. 1990). In the Skatron Argus 100 flow cytometer (Boye et al. 1983; Steen et al. 1990; Boye and Løbner-Olesen 1991) a high-pressure mercury arc lamp is used as the excitation source and this has proved to be more suited to measurements of microorganisms.

It is well documented that the mitochondria of eukaryotic cells have the ability to concentrate "lipophilic" cations such as rhodamine 123 (e.g. Johnson et al. 1980, 1981; Chen et al. 1982; Chen 1988; Grogan and Collins 1990) in an uncoupler-sensitive fashion, and the staining of mitochondria with rhodamine 123, in conjunction with flow cytometry, has been used to study their activity (Darzynkiewicz et al. 1981; Iwagaki et al. 1990; Lizard et al. 1990). Viable bacteria also accumulate rhodamine 123 but non-viable ones do not, and under appropriate conditions the extent to which individ-ual bacteria take up rhodamine 123 quantitatively reflects *the extent of* their viability (Kaprelyants and Kell 1992).

On average, larger cells may be expected to accumulate more molecules of rhodamine 123 than do smaller cells, but since flow cytometry allows collection of both fluorescence (rhodamine 123 uptake) and forward light scattering (cell size) from each cell, the data can thus be plotted as a dual-parameter histogram, enabling one to take size differences between cells into account when interpreting the data.

In contrast to some of the other viability stains (e.g. acridine orange), the uptake of rhodamine 123 not only does not require the use of fixatives to permeabilize the cell, but the concentrative uptake is dependent on an intact and energized cytoplasmic membrane; thus living cells can be used for staining purposes. This has the great advantage that, following staining of the cells, further physiological studies may be conducted if required. Provided that at the concentrations used the dye does not affect the viability of the cells in any way (which we show herein to be true), rhodamine 123 may be expected to be a good probe for assessing the energetic status and/or the viability of gram-positive cells.

Materials and Methods

Organism

Micrococcus luteus (Fleming Strain 2665) was used throughout. It was maintained on nutrient agar slopes at 4°C. The culture was resuscitated in rich liquid medium and streaked onto an agar plate to ensure that the culture was axenic.

Media

The growth medium (lactate minimal medium) used for all continuous culture work was prepared as described in Kaprelyants and Kell (1992). This medium contained (in mg/l): NH_4Cl 4000, KH_2PO_4 1400, biotin 5, L-methionine 20, thiamine 40, inosine 1000, $MgSO_4$ 70, $CuSO_4$ 0.024, $MnCl_2$ 0.5, $FeSO_4$ 1, Na_2MoO_4 0.025, $ZnSO_4$ 0.05, and lithium L-lactate 1000. The pH was adjusted to 7.5 with NaOH prior to autoclaving (121°C, 25 min).

For batch culture a "rich" medium containing 1.3% (w/v) nutrient broth E, pH 7.4 (Lab M), was used.

Carbon-Limited Continuous Culture

Aerobic continuous culture was carried out in an LH Fermentation (Maidenhead, Berks) 500 series fermentor with a working volume of 500 ml. The temperature was controlled at 30°C. The dilution rate was set using a Pharmacia (Bromma, Sweden) Model P1 peristaltic pump connected to a tube that passed below the surface of the broth, ensuring pulse-free addition of fresh medium. The dilution rate was 0.01/h and at least five volume changes were allowed to elapse prior to the removal of samples for flow cytometry.

Flow Cytometry

Flow cytometric experiments were carried out as described previously (Davey et al. 1990; Kaprelyants and Kell 1992) using a Skatron Argus 100 flow cytometer (Skatron Ltd, PO Box 34, Newmarket, Suffolk). The instrument was set up as described in the manual with an additional $0.1 \mu m$ filter placed in the sheath fluid line; sodium azide (1 mmol/l final concentration) was added to the filtered sheath fluid. The sample flow rate was set at 0.5 or $1 \mu l/min$ and a sheath fluid pressure of $1.5-2 kPa/cm^2$ was maintained.

In the present work data were collected from the forward angle (LS1) detector and from a fluorescence detector (FL1) designed for use with fluorescein isothiocyanate (FITC filter). The optical characteristics of this filter are: excitation 470–495 nm, band-stop 510 nm, emission 520–550 nm; these are suitable for use with rhodamine 123.

The flow cytometer was run under the control of a Viglen IIHDE microcomputer (IBM-PC-AT-compatible, 80286 processor, EGA screen), with software supplied by Skatron. This provides data files in a standard format (Murphy and Chused 1984). The photomultiplier voltages were set at 500 and 650 V for the light scattering and fluorescence channels, respectively, and all measurements were gated by the light scattering channel. The forward light scatter and rhodamine fluorescence measured for each cell were "binned" to give histograms of counts versus channel number. An in-house file conversion program written in Microsoft Quick BASIC v4.5 was used to import the data into the spreadsheet program FLOWTOVP. WKS (Davey et al. 1990). Assessment of peak channel numbers was performed on the raw data, whilst data displayed were first subjected to a three-point smoothing routine.

Monodisperse (cv < 2%) latex particles were obtained from the Sigma Chemical Company, Poole, Dorset and from Dyno Particles A/S, PO Box 160, Lillestrøm, Norway.

Preparation of Samples for Flow Cytometry

Samples were removed from the chemostat and placed, in 1 ml aliquots, into Eppendorf tubes. Rhodamine 123 (Sigma) was dissolved in absolute ethanol at a concentration of 1 mmol/l. From this stock solution several dilutions (in ethanol) were made. The amount of rhodamine solution added to each aliquot of cells was kept constant at $1 \mu l$; the addition of $1 \mu l$ ethanol to 1 ml of cells (i.e. 0.1% ethanol) was observed to have no effect on cell viability (data not shown). Cell samples were incubated with the rhodamine 123 for 30–40 min at room temperature prior to analysis by flow cytometry.

Cell Viability Measurements

Standard pour plates consisting of 1.3% Nutrient Broth E (Lab M) solidified with 1.5% purified agar (Lab M) were used to assess cell viability. Serial dilutions of the cell samples from the chemostat were made in sterile phosphate buffer (pH 7.4) at room temperature. Two replicate plates at a suitable dilution (10^{-6} of original sample) were incubated at 30 °C for 4 days and counted manually. To determine whether the addition of rhodamine 123 at the highest concentration used for flow cytometry had any effect on the viability of the cells, rhodamine 123 (1μmol/l final concentration) was added to an identical diluted sample of cells from the chemostat prior to their being plated out. Rich medium was used in the preparation of these plates, but as shown earlier (Kaprelyants and Kell 1992) there was no difference in viabilities seen between these plates and plates consisting of lactate minimal medium solidified with agar.

Total Cell Counts

Total numbers of cells were determined by manually counting unstained cells loaded into an improved Neubauer counting chamber. Because of the small size (less than 1μm) of the cells the counts were performed under oil-immersion at a total magnification of ×1000.

Chemicals

All chemicals were obtained from Sigma or BDH. Water was singly-distilled in all glass apparatus; water used in the sheath fluid of the flow cytometer was treated with a Millipore MilliQ apparatus.

Results

Figure 6.1 shows the light scattering and fluorescence properties of cells of *M. luteus* grown in C-limited continuous culture and prepared and stained with rhodamine 123 as described above. Previous work had shown that under these conditions the uptake of the dye is essentially completely reversed by the addition of the uncoupler carbonyl cyanide *m*-chlorophenylhydrazone (Kaprelyants and Kell 1992). It can be seen here (Fig. 6.1b) that the uptake of rhodamine 123 varies substantially between the individual cells in the population as, to a lesser extent, does the forward light scatter or cell size (Fig. 6.1a). For a given internal concentration, large cells will obviously take up more rhodamine 123 than will small cells, but it is apparent from the dual-parameter histogram (Fig. 6.1c) that even cells of the same size exhibit a broad range of fluorescences. It can also be seen (Fig. 6.1b,c) that under these conditions there is a continuous variation of fluorescence between individuals, from those with very low fluorescences (dead cells) to those with high fluorescence (viable, culturable cells). The cells fluorescing between these extremes will include damaged cells which, given suitable conditions, could be resuscitated and returned to a culturable state (Kaprelyants and Kell 1992). The viability of this particular culture, as judged by a comparison between plate counts and total cell counts, was some 14%. This is somewhat lower than that (*c.* 40%) observed previously for a different run

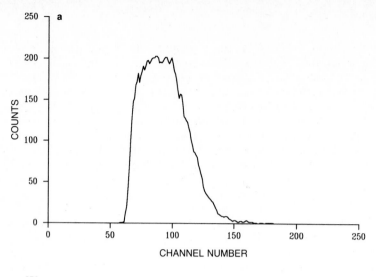

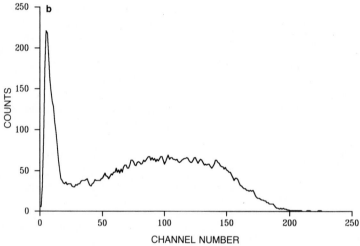

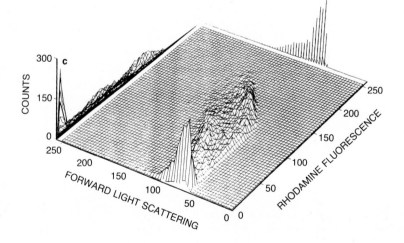

(Kaprelyants and Kell 1992), and may be ascribed to the greater length of time for which this low growth rate had been enforced in the present work.

Figure 6.2 shows the effect of varying the concentration of rhodamine 123 added to the cells prior to flow cytometry. From the lowest concentration of rhodamine used (0.03 μmol/l Fig. 6.2a) up to 0.5 μmol/l (Fig. 6.2d) the peak channel number of the fluorescence histogram increases. Increasing the rhodamine 123 concentration above this level does not further increase the peak channel number of the fluorescence (Fig. 6.2e). With concentrations of rhodamine of 0.5 μmol/l and below there is a good fit by linear regression (correlation coefficient = 0.99) to a plot of the peak channel number of fluorescence versus the rhodamine concentration (Fig. 6.3). In other words, within this linear region the extent to which a cell will take up the dye is, as expected, proportional to the dye concentration. This provides a novel and convenient method for effecting a calibration of the flow cytometer.

From Fig. 6.3 it can be seen that if an average (modal) cell from the population is stained with 0.4 μmol/l rhodamine one would expect it to fluoresce at channel number 131. If the same cell were stained with 0.2 μmol/l rhodamine then it would fluoresce at channel number 102. For any population histogram of rhodamine fluorescence such as those shown in Fig. 6.1b or 6.2a–e, one can therefore say that a cell fluorescing at channel 131 has taken up twice as much rhodamine as a cell fluorescing at channel 102. By repeating this argument for other rhodamine concentrations within the linear region (Fig. 6.3), one can conclude that a difference in peak channel number of 65 units corresponds to a factor of 10 in cell fluorescence. Thus it is possible to obtain a *quantitative* idea of the extent of the heterogeneity of the sample (Fig. 6.4).

Finally, the addition of 1 μmol/l rhodamine 123 to cells prior to plating them out did not affect their viability, since there were, within experimental error, the same number of colonies produced by the cell samples irrespective of whether rhodamine had been added (data not shown).

Discussion

In many microbiological studies it is essential, or at least desirable, to be able to determine the number of viable cells in a sample. Plate counts are the usual way of obtaining viable counts, but the method is slow and does not necessarily give a measure of the number of *viable* cells; more accurately it measures the number of *culturable* cells. Viability staining followed by microscopic examination should give a more accurate measurement but several hundred cells have to be examined for the results to be statistically

---

Fig. 6.1.a–c. Flow cytometry of *Micrococcus luteus* stained with 0.06 μmol/l rhodamine 123. There is heterogeneity in both the forward light scattering (**a**) and the rhodamine 123 fluorescence (**b**) of the sample. However the dual-parameter histogram of forward light scatter and fluorescence (**c**) shows that even cells of comparable size exhibit a broad range of fluorescences (rhodamine uptake).

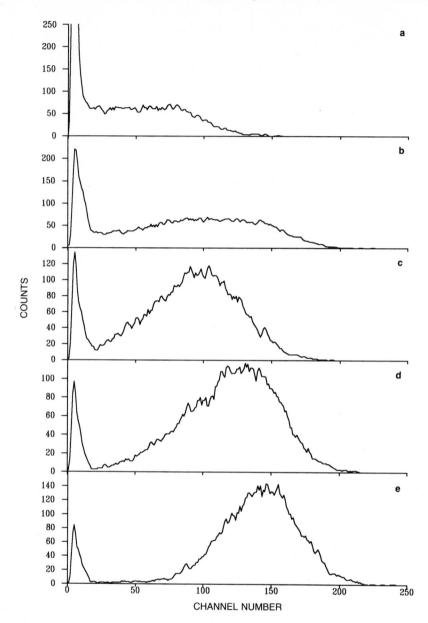

Fig. 6.2.a–e. Increasing the concentration of rhodamine 123 up to 0.5 μmol/l increases the peak channel number of fluorescence (**a–d**). Increasing the rhodamine concentration above 0.5 μmol/l does not increase the fluorescence further (**e**). Peak channel numbers and rhodamine 123 concentration. (μmol/l) for each sample were as follows: **a** 74, 0.03; **b** 87, 0.125; **c** 104, 0.2; **d** 141, 0.5; **e** 141, 1.

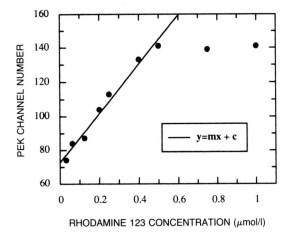

Fig. 6.3. The relationship between rhodamine 123 concentration and the peak channel number of fluorescence is well fitted by linear regression ($r = 0.99$) at concentrations of $0.5\,\mu$mol/l and below. Using higher concentrations of rhodamine 123 does not further increase the fluorescence.

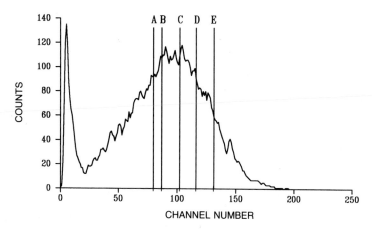

Fig. 6.4. The extent of rhodamine 123 uptake between individual cells is very heterogeneous. Cells were actually stained with $0.2\,\mu$mol/l rhodamine 123. Lines $A-E$ represent the expected fluorescence of an average (modal) cell stained with different concentrations of rhodamine 123. These concentrations (in μmol/l) are as follows: A 0.05, B 0.1, C 0.2, D 0.3, E 0.4. It may be seen that the most fluorescent cells have taken up almost $\times 1000$ the amount of dye taken up by the most weakly fluorescent, but non-dead, cells (at approx. channel 20).

meaningful. As a consequence these tests are tedious to perform, and as such are prone to error.

Cells stained with a suitable viability stain can be studied by flow cyto-metry, a technique that allows rapid acquisition of data. Furthermore, flow cytometry enables judgements to be made on the "degree of viability" (in terms of dye uptake) of *each* cell in the sample, allowing quantification of the heterogeneity of "viability" within the sample. Such a technique has potential for studies of the mechanisms of cell death (Kell et al. 1991; Kaprelyants and Kell 1992). Since, in suitably equipped flow cytometers, multiparameter fluorescence measurements can be made it may be possible to study cell viability in conjunction with other cellular parameters, in order to determine whether they have any role in, or correlation with, cell death.

Since rhodamine 123 does not affect the viability (culturability) of the cells as judged by plate counts, staining of cells does not rule out their use in subsequent physiological studies. In flow cytometers equipped with sorting facilities a fraction of the viable cells could be selected for further study.

We conclude that provided the dye concentration is kept below $0.5\,\mu\mathrm{mol/l}$, flow cytometry in conjunction with rhodamine 123 staining provides a rapid method of assessing cell viability. Furthermore because of the nature of flow cytometric data there are many possibilities for further studies of viability that could not be easily undertaken using any other existing technique.

Acknowledgements

We thank the Science and Engineering Research Council, UK, and the Royal Society, under the terms of the Royal Society/USSR Academy of Sciences exchange agreement, for financial support of this work.

References

Back JP, Kroll RG (1991) The differential fluorescence of bacteria stained with acridine orange, and the effects of heat. J Appl Bacteriol 71:51–58

Boye E, Løbner-Olesen A (1991) Bacterial growth control studied by flow cytometry. Res Microbiol 142:131–135

Boye E, Steen HB, Skarstad K (1983) Flow cytometry of bacteria: a promising tool in experimental and clinical microbiology. J Gen Microbiol 129:973–980

Chen LB (1988) Mitochondrial membrane potential in living cells. Annu Rev Cell Biol 4: 155–181

Chen LB, Summerhayes IC, Johnson LV, Walsh ML, Bernal SD, Lampidis TJ (1982) Probing mitochondria in living cells with rhodamine 123. Cold Spring Harbor Symp Quant Biol 46:141–155

Darzynkiewicz Z, Staiano-Coico L, Melamed MR (1981) Increased mitochondrial uptake of Rhodamine 123 during lymphocyte stimulation. Proc Natl Acad Sci USA 78:2383–2387

Davey CL, Dixon NM, Kell DB (1990) FLOWTOVP: A spreadsheet method for linearizing flow cytometric light-scattering data used in cell sizing. Binary 2:119–125

Dean PN (1990) Data processing. In: Melamed MR, Lindmo T, Mendelsohn ML (eds) Flow cytometry and sorting, 2nd edn. Wiley-Liss, New York

Grogan WM, Collins JM (1990) Guide to flow cytometry methods. Marcel Decker, New York

Harris CM, Kell DB (1985) The estimation of microbial biomass. Biosensors 1:17–84

Hattori T (1988) The viable count: quantitative and environmental aspects. Springer, Berlin

Iwagaki H, Fuchimoto S, Miyake M, Oirta K (1990) Increased mitochondrial uptake of rhodamine 123 during interferon-gamma stimulation in Molt 16 cells. Lymphokine Res 9: 365–369

Johnson LV, Walsh ML, Chen LB (1980) Localization of mitochondria in living cells with rhodamine 123. Proc Natl Acad Sci USA 77:990–994

Johnson LV, Walsh ML, Bockus BJ, Chen LB (1981) Monitoring of relative mitochondrial membrane potential in living cells by fluorescence microscopy. J Cell Biol 88:526–535

Jones RP (1987) Measures of yeast death and deactivation and their meaning. Part 1. Process Biochem 22:118–128

Kaprelyants AS, Kell DB (1992) Rapid assessment of bacterial viability and vitality using rhodamine 123 and flow cytometry. J Appl Bacteriol 72:410–422

Kell DB, Ryder HM, Kaprelyants AS, Westerhoff HV (1991) Quantifying heterogeneity: flow cytometry of bacterial cultures. Ant Van Leeuw 60:145–158

Lizard G, Chardonnet Y, Chignol MC, Thivolet J (1990) Evaluation of mitochondrial content and activity with nonyl-acridine orange and rhodamine 123: flow cytometric analysis and comparison with quantitative morphometry. Cytotechnology 3:179–188

McFeters GA, Singh A, Byun S, Callis PR, Williams S (1991) Acridine orange staining as an index of physiological activity in *Escherichia coli*. J Microbiol Meth 13:87–97

Murphy RF, Chused TM (1984) A proposal for a flow cytometric data file standard. Cytometry 5:553–555

Pollack A, Ciancio G (1990) Cell cycle phase-specific analysis of cell viability using Hoechst 33342 and propidium iodide after ethanol preservation. In: Darzynkiewicz Z, Crissman HA (eds) Flow cytometry. Academic Press, San Diego

Postgate JR (1969) Viable counts and viability. Meth Microbiol 1:611–628

Postgate JR (1976) Death in microbes and macrobes. In: Gray TRG, Postgate JR (eds) The survival of vegetative microbes. Cambridge University Press, Cambridge, pp 1–19

Roszak DB, Colwell RR (1987) Survival strategies of bacteria in the natural environment. Microbiol Rev 51:365–379

Shapiro HM (1988) Practical flow cytometry, 2nd edn. Alan R Liss, New York

Steen HB, Skarstad K, Boye E (1990) DNA measurements of bacteria. In: Darzynkiewicz Z, Crissman HA (eds) Flow cytometry. Academic Press, London

Stoicheva NG, Davey CL, Markx GH, Kell DB (1989) Dielectric spectroscopy: a rapid method for the determination of solvent biocompatibility during biotransformations. Biocatalysis 2:245–255

Characterization of Bacterial Cell Size and Ploidy using Flow Cytometry and Image Analysis

Jerome Durodie, Kenneth Coleman and Michael J. Wilkinson

Introduction

Flow cytometry offers a rapid means of determining the responses of bacteria to a variety of antimicrobial agents each of which has a distinctive mode of action. Characteristics of bacteria, including cell size, DNA and protein content, DNA base composition and immunofluorescence, have been studied by flow cytometry. Although most work has been carried out using *Escherichia coli* (Steen et al. 1982, 1986; Boye et al. 1983; Scheper et al. 1987; Steen 1989; Szabo and Damjanovich 1989; Amann et al. 1990; E. Sahar Tel Aviv University, personal communication), some other bacterial species have been examined, including *Desulphobacter* (Amann et al. 1990), *Pseudomonas*, *Aeromonas*, *Enterococcus* and *Lactobacillus* (Pinder et al. 1990), and a variety of periodontal pathogens (Obernesser et al. 1990). The aims of previous studies have included assessments of the feasibility of detecting bacteria (Amman et al. 1990; Obernesser et al. 1990; E. Sahar Tel Aviv University, personal communication), the identification of bacterial contaminants in food and water (Robertson and Button 1989; Howes 1992) and examinations of the effects of antibacterial compounds on bacterial cell size and DNA content (Steen et al. 1982, 1986; Martinez et al. 1982; Boye et al. 1983; Steen 1990a; von Freesleben and Rasmussen 1991).

Most of these studies used specialized flow cytometers constructed around a mercury vapour arc lamp and specially designed fluidics, epitomized by the equipment designed and built by Steen (1983, 1990b). Such laboratory-built equipment clearly has certain advantages, notably a high level of sensitivity in light scatter detection. At least one commercial version of this type of equipment (the Skatron Argus) became available in recent years and is described elsewhere in this volume.

Our interest is in assessing flow cytometry with respect to: (1) its value in studying the mode of action of antibacterial compounds, and (2) its potential as a rapid screening technique for use with large numbers of samples examined using a variety of protocols.

The results reported here were obtained with an unmodified Becton-Dickinson Facscan equiped with an argon ion 488 nm laser. The Facscan is probably one of the commonest models of flow cytometer in clinical and research use with a potential for routine operation and automatic sample throughput. We have begun by investigating the effects of various β-lactam antibiotics on a single strain, *E. coli* NCTC 10418, which we characterized with regard to DNA ploidy (by fluorescence) and cell size (by forward scatter).

Bacterial DNA Synthesis

The study of bacterial DNA poses certain problems compared with that of mammalian cell lines. In eukaryotes, DNA replication follows a regular, defined cycle of replication, chromosome separation and rest. This means

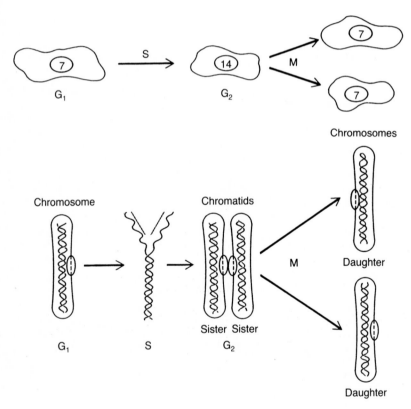

Fig. 7.1. The eukaryotic cell cycle (in *Dictyostelium discoideum*). G_1 phase, each nucleus has a full complement of 7 chromosomes; S phase, the chromosomes are all replicating; G_2 phase, each nucleus has twice the number of chromosomes (14); M phase, one copy of each chromosome goes to each daughter cell. (Reproduced from U. Goodenough, *Genetics*, 2nd edn, with permission of the publishers Holt-Saunders International.)

that single chromosomes and their multiples can be observed easily (Fig. 7.1). Bacterial DNA replication, however, is analogous to a continuous rolling circle. As soon as the replication fork is sufficiently distant from the origin of replication, a new replicative fork may begin (Fig. 7.2) if the rate of cell division is sufficiently high to merit it. Any complete circle of DNA may contain many variable-sized fragments of replicating DNA associated with it, prior to a full duplicate version separating off. A snapshot of bacterial DNA will consequently always represent a gross DNA content and not discrete unit values of DNA at specific time points, as is the case in eukaryotes.

One approach to measuring the DNA content and ploidy of bacterial cells is pre-treatment with a protein synthesis inhibitor. Three such inhibitors have been used: rifampicin (Boye and Løbner-Olesen 1991; von Freesleben and Rasmussen 1991), chloramphenicol (Boye et al. 1983; Steen 1989) and olivomycin (Shapiro 1988). Rifampicin, although primarily a protein

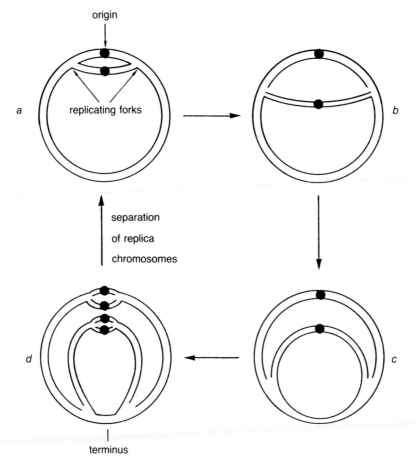

Fig. 7.2. Prokaryotic DNA replication. (Reproduced from B. Lewin, *Gene expression*, vol. 1, with permission of the publishers John Wiley & Sons.)

synthesis inhibitor, inhibits DNA synthesis by preventing the initiation of new replicative loops. Any DNA replication already begun before the addition of rifampicin will continue to completion and then stop (Kersten and Kersten 1974). Thus, the presence of many variable-sized fragments of DNA is eliminated: the bacterial chromosome will comprise "whole numbers" of DNA strands. The bacterial population will show discrete DNA peaks reflecting whole numbers of chromosomes. Pre-treatment with rifampicin thus makes it possible to examine the DNA ploidy of a bacterial population and the effects of treatment with antibacterial compounds on DNA synthesis.

The nucleic acid intercalating dyes, notably propidium iodide, used to stain bacterial DNA for analysis in argon ion laser instruments also intercalate with RNA. This is particularly significant because bacteria contain a higher proportion of RNA per cell than do eukaryotes. It is generally recognized that some form of RNAse treatment of cells is necessary prior to analysis, but the effectiveness of such treatment is disputed. Some workers claim that this is not effective (Robertson and Button 1989); others suggest that the problem has proved impossible to resolve (Steen 1990a) or that the results obtained after RNAse treatment are not as consistent as those obtained using alternative DNA-specific stains requiring ultraviolet (u.v.) excitation (Steen 1990b). In our experience, treatment with RNAse of ethanol-fixed cells improves the clarity of DNA detection significantly and works consistently. Treatment of *E. coli* with DNAse alone leaves a low level of undifferentiated RNA-associated fluorescence, and treatment with both enzymes removes detectable fluorescence altogether when cells are then stained with propidium iodide.

DNA and cell size analysis after various antibiotic treatments are also documented in the literature (Steen et al. 1982, 1986; Martinez et al. 1982; Boye et al. 1983; E. Sahar personal communication). Although the particular compounds we have used have not been studied previously, similar β-lactams with comparable effects have been investigated. As the first stage in assessing the value of flow cytometry for an antimicrobial research programme and for screening purposes, we have explored the extent to which DNA content and cell size can be characterized following treatment with well-known antibiotics.

Preparation of Bacteria for Flow Cytometry

A simple protocol for the treatment and preparation of *E. coli* cultures is shown in Fig. 7.3. An overnight broth culture is diluted into fresh, filtered, pre-warmed broth and incubated at 37 °C aerobically in a shaking water-bath (250 rpm) until it has entered early log phase. Antibiotic is added and samples are taken at regular intervals. Duplicate 5.0 ml samples are pelleted by centrifugating at $1500-2000\,g$ for 5 min. One aliquot is resuspended in an equal volume of fresh, pre-warmed broth with addition of $150\,\mu g/ml$ rifampicin, and is incubated at 37 °C for a further 2 h, when the cells are again pelleted. The second aliquot is left untreated. Both rifampicin-treated and non-treated cell pellets are resuspended in ice-cold Tris-HCl buffer pH

Overnight _E. coli_ culture

↓

t = 0
Diluted 10⁻³ into fresh, pre-warmed broth

↓

t = 2
Add test antibiotic

Samples taken @ t = 0, 2 and hourly thereafter.

SAMPLE PREPARATION:

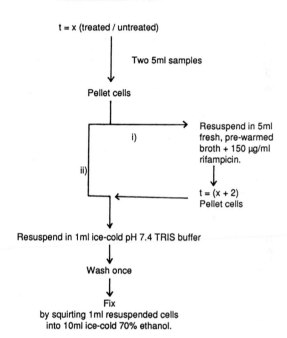

t = x (treated / untreated)

↓ Two 5ml samples

Pellet cells

i) → Resuspend in 5ml fresh, pre-warmed broth + 150 µg/ml rifampicin.

↓

t = (x + 2)
Pellet cells

ii)

↓

Resuspend in 1ml ice-cold pH 7.4 TRIS buffer

↓

Wash once

↓

Fix
by squirting 1ml resuspended cells
into 10ml ice-cold 70% ethanol.

PREPARATION FOR FLOW CYTOMETRY ANALYSIS:

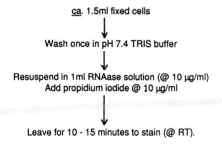

ca. 1.5ml fixed cells

↓

Wash once in pH 7.4 TRIS buffer

↓

Resuspend in 1ml RNAase solution (@ 10 µg/ml)
Add propidium iodide @ 10 µg/ml

↓

Leave for 10 - 15 minutes to stain (@ RT).

Fig. 7.3. Protocol for the preparation of bacterial samples for flow cytometric analysis.

7.4, washed once in the same and fixed by squirting forcefully from a Pasteur pipette into 10 ml ice-cold 70% ethanol to ensure rapid dispersal of cells. Cells fixed in this manner can be stored at 4°C for several months without deterioration before staining and analysis.

Propidium iodide has an excitation maximum at 495 nm and an emission maximum at 639 nm and is well suited for use in an argon ion laser based system. The combination dye ethidium bromide–mithramycin A, which requires u.v. excitation, has been used by ourselves and other workers when operating mercury arc lamp instruments. The excitation and emission wavelengths of this combination, however, are unsuitable for argon ion laser instruments. Propidium iodide, and ethidium bromide–mithramycin A, can only penetrate bacterial cells that have been permeabilized, and this is achieved by fixation in 70% ethanol.

To stain bacteria for flow cytometry a small volume (c. 1.5 ml) of fixed cells are pelleted, washed in pH 7.4 Tris buffer, and resuspended in 1 ml RNAse solution. Propidium iodide solution (100 μg/ml) is added to a final concentration of 10 μg/ml and the mixture incubated for 15 min at room temperature before the cells are analysed. Any subsequent dilution of the cell sample during analysis is carried out with RNAse-stain solution to avoid leaching stain from cells. Non-treated controls are prepared in the same way, but with the omission of the antibiotic.

Following the above procedure, cells treated with two different β-lactam antibiotics at sub-inhibitory concentrations (one-half or one-quarter of the minimum inhibitory concentrations, MICs) have been analysed. The antibiotics used have well-reported modes of action, both of them binding to specific penicillin binding proteins (PBPs): amoxycillin binding primarily to PBP1 (Curtis et al. 1979; Matsuhashi et al. 1981; Lorian 1986) and mezlocillin to PBP3 (Spratt 1975, 1977; Lorian 1986).

Operation of the Flow Cytometer

An unmodified Becton-Dickinson Facscan instrument was used in this work. It is powered by a 15 mW, 488 nm argon ion laser. This intercepts the sample stream via a quartz flow cell. The forward light scatter signal is detected by a photodiode. The four other detectors (one for side scatter plus three fluorescence detectors) are all photomultipliers. A Hewlett Packard computer and operating system is used to run the data display and analysis software supplied by Becton-Dickinson. The sheath fluid used was "Facsflow" (Becton-Dickinson), a commercial preparation of phosphate-buffered saline. A sample flow rate of 12 ml/min was used in all bacterial work. Particular care was taken to ensure good optical alignment of the system and cleanliness of the optical components to optimize the signal-to-noise ratio. When assessing bacteria rather than mammalian cells on this instrument, we found that optical alignment is more critical and requires more regular attention.

Turbulence can be a significant problem in bacterial work because it can cause the sample stream to be diverted away from the centre of the laser beam and may also interfere with the smooth passage of the cells as a linear stream through the flow cell. The presence and/or build-up of particulates must also be minimized to avoid generating unwanted noise in the vicinity of the bacterial signals for scatter and fluorescence. We have therefore adopted several additional measures to avoid the formation of turbulence and to minimize contamination in the fluidics:

1. Elimination of air bubbles in the tubing, filter and flow cell by draining and filling the flow cell several times to purge air and ensure even wetting of internal surfaces.
2. Avoidance of factors that might cause local perturbations in fluid flow such as kinks in tubing.
3. Maintenance of a low to moderate count rate (typically 300 cells/s) to avoid cell–cell interactions that may impede linear flow of the sample.
4. Use of pre-filtered solutions in the sample preparation to minimize contamination of the system with particulates.
5. Regular cleaning of tubing with a detergent after each session to prevent the build-up of debris and the formation of microbial contaminants.
6. Replacement of components showing marginal defects such as tubing that is worn or roughened, or a flow cell having scratches in the optical path, to routinely achieve a high standard of maintenance.

In mammalian cell work it is customary and usually very straightforward to set up the flow cytometer and check its performance using some form of beads, such as fluorescent latex microspheres. Some workers claim to have found these reliable for bacterial work (Amman et al. 1990; Obernesser et al. 1990), while others have found them unacceptable (Pinder et al. 1990). Alternatives such as staphylococci (Robertson and Button 1989) have been investigated. Our experience of beads is that they have only limited value in preparing the instrument for use in analysing *E. coli* samples, since the light scattering and fluorescent properties of beads (size range $1-2\,\mu m$) are distinctly different from those of bacterial rods of similar size. The most useful test material is a bacterial sample, and we keep in long-term store a large quantity of ethanol-fixed *E. coli* for use as quality controls.

An additional convenience when examining relatively large eukaryotic cells is that the discrimination threshold can be set at a level that will eliminate all signal and noise in the lower channels since these do not generally contain important data. This is not so in bacterial work and care must be taken to select a threshold setting that removes true noise in the lowest channels while not interfering unduly with the collection of data from channels just above these. We have found that a threshold discrimination setting of approximately 30 is usually satisfactory and this is checked using quality control samples of bacteria.

Results

The difference between rifampicin-treated and untreated cells when analysed by flow cytometry can be seen for an early log phase culture of *E. coli* in Fig. 7.4. Without rifampicin, a single, broad DNA peak is observed indicat-

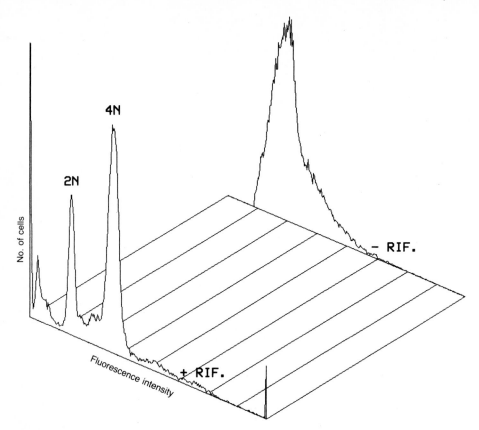

Fig. 7.4. The effect of rifampicin treatment on DNA peak appearance in *E. coli*. (Note that the fluorescence scale is linear.)

ing the wide range of DNA content present in the population. The same sample pre-treated with rifampicin produces two distinct DNA peaks. These two DNA peaks represent the population of cells containing two chromosomes (*2n*) and those containing four chromosomes (*4n*). The channel number displayed along the *x*-axis of Fig. 7.4 is a measure of fluorescence intensity. The median value channel for the *4n* cells was seen to be double that of the *2n* cells, as would be expected (in the results shown, these were 185 and 367 respectively). This result also indicates that the fluorescence detector has a satisfactorily linear response even across the relatively low-intensity channels used in bacterial work.

The coefficients of variation (c.v.) of these fluorescence peaks were 7.1% and 6.3% respectively. The c.v. are therefore not as low as is achievable with alternative instruments specially designed for bacterial flow cytometry, where figures as low as 2.5% have been quoted (Steen 1990b). Nonetheless, they are good enough to allow clear resolution of distinct DNA peaks in this type of experiment.

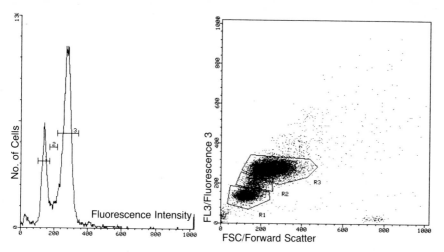

Fig. 7.5. Sub-populations of a control *E. coli* sample with respect to both DNA ploidy and cell size: R1, 2*n* cells; R2, 3*n* cells; R3, 4*n* cells.

Forward scatter is an approximate measure of cell size, and a dot-plot of fluorescence intensity (DNA content) versus forward scatter can be used to compare the physical size of cells in the 4*n* and 2*n* peaks (Fig. 7.5). The 2*n* cells are nearer the origin of the forward scatter axis, i.e. smaller, than their 4*n* counterparts.

The effects of a half MIC concentration of amoxycillin (0.5 µg/ml) on the DNA content of treated *E. coli* NCTC 10418 are shown in Fig. 7.6. The DNA peaks shift to the left with continuing exposure to the antibiotic. The untreated 2*n* and 4*n* peak positions are shown in the foreground. After 2 h of treatment the 4*n* component of the population has virtually disappeared, leaving 2*n* and 1*n* cells. After 5 h the population consists almost solely of 1*n* cells.

Amoxycillin binds preferentially to PBP1, an enzyme complex responsible for synthesis of new cell wall material. Inhibition of PBP1 weakens the cell wall and leads to lysis. This explains the presence of a fluorescence peak below that of the monoploid (1*n*) cells at 2 h and at 5 h that is probably due to DNA fragments released into the medium by lysed cells (Fig. 7.6). The progressive shift in DNA peaks observed by flow cytometry indicates that DNA replication is slowed down prior to cell death. It is known that cell wall synthesis and DNA replication are intimately linked (Lorian 1986), and it is likely that a reduction in the rate of murein synthesis would be related to a slowing down of the rate at which fork-initiation occurs. Consequently, as treatment progresses there are fewer of the fast-dividing 4*n* cells, then fewer 2*n* cells and finally a majority of 1*n* cells, i.e. cells which are non-dividing or very slow growing but as yet unlysed.

Figure 7.7 shows the effects of a quarter MIC of mezlocillin (0.06 µg/ml) on the DNA content of *E. coli* NCTC 10418. The expected 2*n* and 4*n* peak positions are shown in the foreground. The absolute position of these peaks is different from that in the experiment described above because the

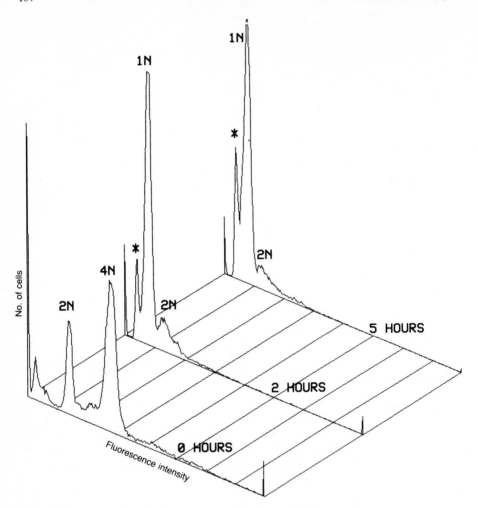

Fig. 7.6. DNA ploidy effects in *E. coli* treated with one-half MIC amoxycillin. *Asterisks* indicate free DNA. (Note that the fluorescence scale is linear.)

settings (detector and amplifier gain) of the instrument have been adjusted in order to allow for the expected increase in cell size resulting from this treatment. After 1 h of treatment the DNA peak is equivalent to a ploidy level of $15n$ to $20n$. Mezlocillin inhibits PBP3 of *E. coli*, a protein known to be involved in cross-wall formation during cell division (Spratt 1977). Inhibition of PBP3 results in filamentation, where cells continue to elongate without septation. These filaments often extend to 4–16 times the normal cell length and eventually die.

This shift in both cell size and DNA ploidy can also be seen on dot-plot analysis of these results. Fig. 7.8 shows such a comparison for mezlocillin-treated cells and controls. In comparison with the controls (Fig. 7.8a), the

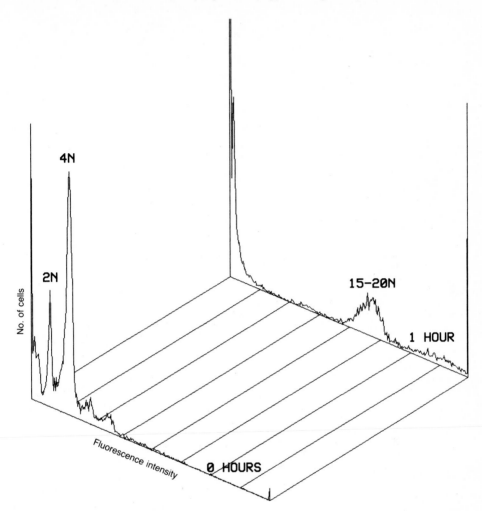

Fig. 7.7. DNA ploidy effects in *E. coli* treated with one-quarter MIC mezlocillin. (Note that the fluorescence scale is linear.)

mezlocillin-treated cells (Fig. 7.8b) show a higher fluorescence emission and have a wider range of forward scatter signals. Region 1 is background noise.

Measurement of Bacterial Cell Size by Light Microscopy and Image Analysis

The cell size data obtained from flow cytometry studies were compared with data obtained on parallel samples analysed by light microscopy and image analysis.

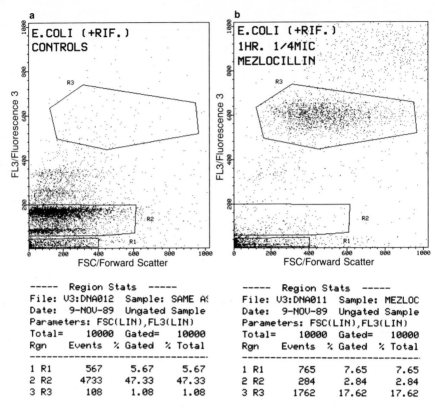

Fig. 7.8. a Cell size and DNA ploidy sub-populations in a control *E. coli* culture. R1, background noise; R2, *2n* and *4n* cells of normal size; R3, the longer 15–20*n* cells. **b** The effect of one-quarter MIC mezlocillin treatment on both DNA ploidy and cell size of *E. coli*.

A primary requirement when analysing rod-shaped bacteria such as *E. coli* by light microscopy is that the long axis of all cells be presented at right angles to the optical axis of the microscope and within a narrow focal plane. In practice this means ensuring that the cells are attached lengthways to the glass slide as a monolayer. This was achieved by centrifuging bacterial suspensions onto a glass slide at $300\,g$ for 10 min using a Cytospin (Shandon Southern Ltd, Runcorn, Cheshire, UK). Cells were imaged with a Zeiss Axioplan microscope using differential interference contrast (Normarski) optics with a ×63/1.4 NA oil-immersion objective lens. Cell length was measured using a commercial video image analysis system based on a 386 16 Mz IBM-compatible PC ("Freelance", Sight Systems, Newbury, Berkshire, UK). The only operator intervention required was initial calibration with a graticule and then marking the ends of each cell with the on-screen cursor prior to automatic measurement.

Table 7.1 compares cell size measurements by image analysis and flow cytometry. In all cases the figures are relative to control cells of unit length.

Table 7.1. A comparison of *E. coli* NCTC 10418 cell size by image analysis and flow cytometry

Treatment	Mean cell size measured by:		Flow cytometry (relative value)
	Microscopy		
	Absolute value (μm)	Relative value	
Control	3.3 ± 0.6	1.0	1.0
Amoxycillin ($\frac{1}{2}$ MIC at 2 h)	2.1 ± 0.9	0.6	0.5
Mezlocillin ($\frac{1}{4}$ MIC at 1 h)	15.1 ± 9.6	4.6	5.0

Table 7.2. Cell size and ploidy (*n*) of *E. coli* NCTC 10418

Treatment	Relative cell size	Chromosome number (% of population)			
		$1n$	$2n$	$3n$	$4n$
Control	1.0	–	25	14	50
Amoxycillin ($\frac{1}{2}$ MIC at 2 h)	0.5	31	57	6	4
Mezlocillin ($\frac{1}{4}$ MIC at 1 h)	5.0	*c.* $17n$: 70%			

Forward scatter on the flow cytometer compares very favourably with image analysis for studies of cell size changes following treatment with amoxycillin and mezlocillin. Flow cytometry is a far more efficient means of collecting such data: the data shown here are based on 10 000 cells for each flow cytometry measurement (acquired in a matter of seconds) but only 100 cells for each direct microscopy measurement (acquired after several hours). However, it should be borne in mind that microscopy and image analysis potentially give very accurate absolute values for cell size in studies such as this because calibration standards are readily available. Obtaining similar absolute values for bacterial samples by flow cytometry is not so straightforward, given the paucity of calibration specimens that have scatter characteristics close to those of bacterial cells.

An analysis of sub-populations with respect to chromosome number and drug treatment is summarized in Table 7.2. This shows that:

1. The control cells, regarded as having unit length for comparative purposes, consist substantially of $2n$ and $4n$ cells representing 25% and 50% of the total population.

2. Two hours of one-half MIC amoxycillin treatment gives a reduction in cell size of approximately one-half, and such a population consists substantially of $1n$ and $2n$ cells, representing 31% and 57% of the total.

3. One hour of one-quarter MIC mezlocillin treatment results in filamentation, with 70% of the population displaying a ploidy of approximately $17n$.

Conclusions

These preliminary studies demonstrate that relative changes in bacterial cell size (ranging from reduction and lysis to filamentation), and the associated changes in cell cycle and DNA ploidy, caused by treatment with various β-lactam antibiotics are readily detectable with a conventional laser-based flow cytometer. The drug concentrations at which such effects can be easily and unambiguously detected by flow cytometry are sub-lethal, and it should be remembered that some traditional microbiological tests would not readily detect such changes at these low concentrations. Furthermore, the results relating to cell size obtainable by flow cytometry are potentially available in a relatively brief time compared with the more labour-intensive techniques involved in image analysis. However, the view that flow cytometry of bacteria, using any design of instrument, can always provide results rapidly and routinely is tempered to some extent by the practicalities of operating a system that has relatively complex optics and fluidics under computer control. The experience of ourselves and other workers with a variety of flow cytometry systems is that maintenance of the hardware and software involves a significant investment in time and resources that is generally greater than might be necessary for an image analysis system linked to a microscope.

The value of flow cytometry for observing the effects of antimicrobial action on gram-negative rods is evident. The potential for extending such studies within the field of antimicrobial research is very exciting. A wide range of fluorescent probes is presently available for estimating cellular parameters (other than nucleic acid content) such as protein content, enzyme activity, calcium flux, membrane potential and intracellular pH. The challenge in applying flow cytometry more broadly in bacterial and pharmaceutical research lies in: (1) developing the methodology of specimen preparation to exploit the potential available from this range of dyes, and (2) improving the instrumentation needed for bacterial work, by optimizing and, where necessary, modifying existing equipment. The potential of flow cytometry for investigating drug efficacy, for elucidating the mode of action of drugs and for screening new compounds for antibiotic activity justifies further effort in this emerging field.

Acknowledgements

We would like to thank Dr. R. Jepras (Centre for Applied Microbiology and Research, Porton Down, Wiltshire) for his assistance in our early endeavours in bacterial flow cytometry, and staff at Becton-Dickinson for their helpful cooperation.

References

Amann RI, Binder BJ, Olson RJ, Chisholm SW, Devereux R, Stahl DA (1990) Combination of 16S rRNA-targeted oligonucleotide probes with flow cytometry for analysing mixed microbial populations. Appl Environ Microbiol 56:1919–1925

Boye E, Løbner-Olesen A (1991) Bacterial growth control studied by flow cytometry. Res Microbiol 142:131–135

Boye E, Steen HB, Skarstad K (1983) Flow cytometry of bacteria: a promising tool in experimental and clinical microbiology. J Gen Microbiol 129:973–980

Curtis NAC, Orr D, Ross GW, Boulton MG (1979) Affinities of penicillins and cephalosporins for the penicillin binding proteins of Escherichia coli K-12 and their antibacterial activity. Antimicrob Agents Chemother 16:533–539

Howes G (1992) Stepping ahead: applications for flow cytometry now go beyond the usual analysis of lymphocyte subpopulations. Lab News, February 19

Kersten H, Kersten W (1974) Inhibitors of nucleic acid synthesis. In: Kleinzeller A, Springer G, Whittmann H (eds) Molecular biology, biochemistry and biophysics. Springer, Berlin, pp 107–117

Lorian (ed) (1986) Antibiotics in laboratory medicine, 2nd edn. Williams and Wilkins, Baltimore

Martinez OV, Gratzner HG, Malinin TI, Ingram M (1982) The effect of some β-lactam antibiotics on Escherichia coli studied by flow cytometry. Cytometry 3:129–133

Matsuhashi MF, Ishino F, Nakagawa J, Mitsui K, Nakajima-Iijima S, Tamaki S (1981) Enzymatic activities of penicillin binding proteins of Escherichia coli and their sensitivities to β-lactam antibiotics. In: Salton M, Shockman G (eds) β-lactam antibiotics: mode of action, new development and future prospects, Academic Press, New York, pp 169–184

Obernesser MS, Socransky SS, Stashenko P (1990) Limit of resolution of flow cytometry for the detection of selected bacterial species. J Dental Res 69:1592–1598

Pinder AC, Purdy PW, Poulter SAG, Clark DC (1990) Validation of flow cytometry for rapid enumeration of bacterial concentrations in pure cultures. J Appl Bacteriol 69:92–100

Robertson BR, Button DK (1989) Characterising aquatic bacteria according to population cell size, and apparent DNA content by flow cytometry. Cytometry 10:70–76

Scheper T, Hitzman B, Rinas U, Schügerl K (1987) Flow cytometry of Escherichia coli for process monitoring. J Biotechnol 5:139–148

Shapiro HM (1988) In: (ed) Practical flow cytometry, 2nd edn. Alan R. Liss, New York

Spratt BG (1975) Distinct penicillin binding proteins involved in the division, elongation and shape of Escherichia coli K-12. Proc Natl Acad Sci USA 72:2999–3003

Spratt BG (1977) Properties of the penicillin binding proteins of Escherichia coli K-12. Eur J Biochem 72:341–452

Steen HB (1983) A microscope based flow cytophotometer. Histochem J 15:147–160

Steen HB (1989) Flow cytometry in the pharmaceutical industry. In: Barber M (ed) Pharmaceutical manufacturing international. Sterling Publications, pp 89–94

Steen HB (1990a) Flow cytometric studies of micro-organisms. In: Melamed M, Lindmo T, Mendelsohn M (eds) Flow cytometry and sorting, 2nd edn. Wiley-Liss, New York, pp 605–622

Steen HB (1990b) DNA measurements of bacteria. In: Darzynkiewicz Z, Crissman HA (eds) Methods in cell biology, 33. Academic Press, London, pp 519–526.

Steen HB, Boye E, Skarstad K, Bloom B, Godal T, Mustafa S (1982) Applications of flow cytometry on bacteria: cell cycle kinetics, drug effects and quantitation of antibody binding. Cytometry 2:249–257

Steen HB, Skarstad K, Boye E (1986) Flow cytometry of bacteria: cell cycle kinetics of antibiotics. Ann N Y Acad Sci 468:329–338

Szàbo G Jr, Damjanovich S (1989) Fluorescent micrococci as microbeads. Cytometry 10: 801–802

von Freesleben U, Rasmussen KV (1991) DNA replication in Escherichia coli gyrB(Ts) mutants analysed by flow cytometry. Res Microbiol 142:223–227

Chapter 8

Cytometric Evaluation of Antifungal Agents

Elizabeth A. Carter, Frank E. Paul and Pamela A. Hunter

Introduction

The use of flow cytometry has revolutionized the study of immunology such that it is difficult now to envisage a modern laboratory without access to a cytometer. Other disciplines, such as haematology, are catching up. In contrast the technique has had limited impact on microbiology and most microbiological studies, to date, have been with bacteria (Shapiro 1990). The question: What advantages does the adoption of expensive technology offer and what are the shortcomings of conventional microbiological assays? We will address this issue with reference to testing the efficacy of antifungal agents against *Candida albicans*.

The use of flow cytometry to evaluate effects of antifungal agents of fungal cells is still in its infancy (Scott-Pore 1990; O'Gorman and Hopfer 1991), yet the technique could be a useful alternative to conventional tests. Traditional methods have a number of drawbacks. They mostly depend on the visual determination of a growth/no growth end point in broth cultures or on agar containing a series of concentrations of the drug, this marking the MIC – the lowest concentration of the drug that inhibits growth of the fungus completely, or almost completely. Thus the tests take the time required for visible growth to occur and so usually include at least an overnight incubation step. Also, without refinements or additional tests, they yield little information apart from the MIC. It is not obvious whether the drug is fungicidal or merely fungistatic; whether it has effects at sub-MIC concentrations, for example, on morphology; whether the whole population is equally susceptible. All these factors could have a bearing on the therapeutic efficacy of a drug. The tests do not give any indication of the drug's mode of action either. Flow cytometry has the potential to provide more information, more rapidly, on individual cells as well as on the whole population. It can also reveal immediate or short-term effects on specific cellular processes, as well as longer-term, more gross effects, including effects on cell viability.

This depends to a large extent on the feasibility of using fluorescent dyes for assessing cell properties. Many fluorescent probes are now available for work with mammalian cells and some of these should find use in the evaluation of antifungal agents. Other probes, such as ChemChrome Y from Chemunex, are being developed specifically for use with fungi for applications in the food and brewing industries. Other dyes that have an affinity for components of fungal cell walls may also find applications in flow cytometry, e.g. calcofluor white for chitin and aniline blue for glucan.

Our own work has so far concentrated on finding suitable probes to monitor the interaction between amphotericin B and *Candida albicans*. Amphotericin B has been in clinical use for over 30 years and, despite the advent of newer agents, is still regarded as the "gold standard" of antifungal chemotherapy (Gallis et al. 1990). Its mode of action is not completely understood (Brajtburg et al. 1990). It is generally accepted that it interacts with sterols in cell membranes, thereby increasing membrane permeability, but whether the consequent loss of ions and essential metabolites kills the organism is questioned. Oxidative damage to the membrane has been proposed as a second, fungicidal mode of action (Sokol-Anderson et al. 1986). Amphotericin B acts on mammalian cells as well as on fungal cells and it is thought that it can only be used clinically because it has a greater affinity for, or a greater effect on, the ergosterol-containing membranes of fungi than on the cholesterol-containing membranes of mammalian cells (Brajtburg et al. 1974). Flow cytometry can be used to investigate its action on mammalian cells as well as on fungal cells, as demonstrated in a recent paper (Jullien et al. 1991) reporting on the use of a membrane potential dye for monitoring effects of amphotericin B on leukocytes.

We chose *Candida albicans* as our target organism for technical reasons and because it is a clinically relevant organism.

Table 8.1. Probes tried

Probe	Activity[a]
Probes developed for use with mammalian cells	
Propidium iodide	A marker for membrane integrity, excluded by intact cells, taken up by cells that have lost the integrity of their cell membrane (Shapiro 1988)
Fluorescein diacetate	A viability marker, taken up and cleaved to fluorescein by viable cells (Shapiro 1988)
Indo-1, Fluo-3, MagIndo, PBF1 and SB1	Probes sensitive to intracellular concentrations of ions, taken up as fluorogenic substrates, cleaved to ion-sensitive product by enzymes in cell (Molecular Probes Technical Brochure)
Oxonol	A dye sensitive to membrane potential, partitioning into mammalian cell membranes to an extent that depends inversely on membrane potential. Thus viable cells are lightly stained, while dying cells that have lost the ability to maintain a membrane potential stain brighter (Wilson and Chused 1985)
Probes developed for use with fungal cells	
ChemChrome Y	According to suppliers, Chemunex, non-fluorescent substrate is taken up by viable cells and cleaved by enzymes involved in ergosterol biosynthesis to fluorochrome

[a] In mammalian or fungal cells according to the use for which the probe was developed.

Probes

We have looked at a number of probes, most of which have generally been used with mammalian cells (Table 8.1). There are few reports (Scott-Pore 1990) of any having been used in flow cytometric studies of fungi, though some have been used in fluorescence microscopy of fungi (Butt et al. 1989). All probes were from Molecular Probes, except ChemChrome Y which was from Chemunex.

Approaches

The ability of the probes to stain *C. albicans*, treated or untreated with amphotericin B as appropriate, was examined. They were then selected for further evaluation depending on the results. For more detailed studies of selected probes we adopted several approaches. In one series of experiments to investigate the correlation between viability and staining of the cells with the fluorescent dyes, cell suspensions in phosphate-buffered saline, 10^6 cells/ml, were treated overnight at 37 °C with a range of concentrations of amphotericin B, and probes were added the following day. In a second series of experiments the probes were added to the cell suspensions at the same time as amphotericin B to assess their usefulness in following the more immediate effects of the antibiotic on the cells. With this approach it is obviously important to check, as far as possible, that the probes do not influence the interaction between the antibiotic and the cells, that they neither reduce nor exacerbate the antifungal effects. Thus probes were added to conventional microtitre MIC tests, and added to cultures in a biophotometer that measures growth turbidometrically with time (Coleman et al. 1983) to see whether they did affect the activity of amphotericin B.

We use a "Facstar plus" from Becton-Dickinson modified with a quartz flow cell, and Lysis II software, also from Becton-Dickinson, for data acquisition and analysis, but almost any commercially available flow cytometer should be suitable. All probes were excited at 488 nm and fluorescence measured using a 530/30 dichroic filter, except for fluorescence from propidium iodide which was measured with a 630/22 dichroic filter. Fluorescence is expressed in arbitrary units and mean fluorescence data have been normalized according to the fluorescence of control cells.

Results

In early experiments fluorescein diacetate (used at 0.5μg/ml) was readily taken up and cleaved to fluorescein by untreated yeast cells, but leaked out too readily to be useful. Also when the cells were examined microscopically the remaining dye appeared to be sequestered in cell organelles. Indo-1 and the other probes used for monitoring intracellular concentrations of ions in mammalian cells (tested at $1-5 \mu$mol/l) were apparently not taken up by

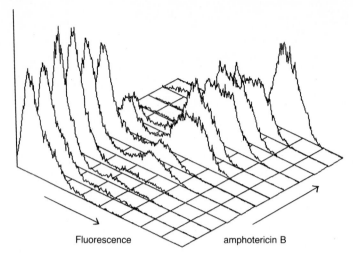

Fig. 8.1. Fluorescence of yeast cells treated overnight with amphotericin B (left to right: 0, 0.01, 0.02, 0.05, 0.1, 0.2, 0.5, 1, 5, 10 and 100 μg/ml) and of cells killed in a boiling water bath, then labelled with propidium iodide (1 mg/ml) for 3 h.

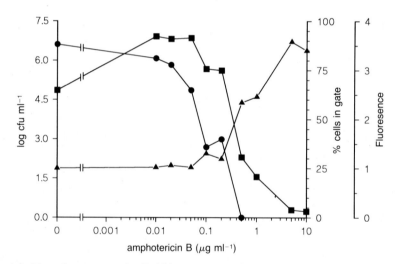

Fig. 8.2. Mean fluorescence of cells (▲), percentage of cells falling within a low fluorescence gate (■) and numbers of viable cells (●) in cell suspensions treated overnight with amphotericin B then labelled with propidium iodide (1 μg/ml) for 3 h. Fluorescence data have been normalized according to fluorescence of untreated cells.

the yeast cells. Only propidium iodide, ChemChrome Y and oxonol were studied in more detail.

Propidium Iodide

Propidium iodide (used at 1 μg/ml) was excluded from viable yeast cells, as from viable mammalian cells, but was taken up by dead yeast cells, staining

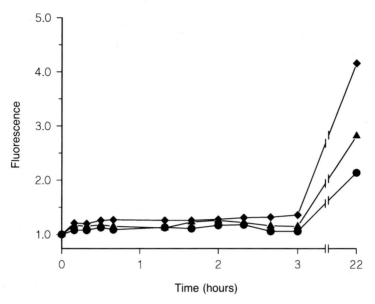

Fig. 8.3. Fluorescence of cells treated with amphotericin B and labelled with propidium iodide. Amphotericin B (0, ●; 1, ▲; and 10, ◆ μg/ml) and propidium iodide (1 μg/ml) added to cell suspensions at time 0. Fluorescence data normalized according to fluorescence at time 0.

them a bright red. When yeast cells were treated overnight with amphotericin B then labelled with propidium iodide, two distinct populations of cells, distinguishable by their fluorescent intensity, were evident (Fig. 8.1). The proportion of cells in the more fluorescent population increased with the concentration of amphotericin B used above a threshold of 0.5 μg/ml; thus the mean fluorescence of the total cell population increased in a dose-dependent manner with amphotericin B, and the percentage of cells with only low fluorescence falling within a certain gate decreased along with the viable count of the cell suspension, also in a dose-dependent manner, for concentrations of amphotericin B of 0.5 μg/ml and above (Fig. 8.2). However, when propidium iodide was added to cell suspensions at the same time as amphotericin B, no change in the fluorescence of the cells was apparent until after prolonged incubation (Fig. 8.3). Thus propidium iodide provides a good measure of cell death but is not a very sensitive probe for monitoring the effects of amphotericin B on yeast cells.

ChemChrome Y

In contrast to propidium iodide, Chemchrome Y (used at dilution of 1/5000) was rapidly taken up and cleaved to its fluorochrome by untreated yeast cells (within minutes), as shown by a rapid increase in their fluorescence. However, it was taken up less rapidly by cells treated overnight with amphotericin B, even with concentrations as low as 0.01 μg/ml, and not at all by cells that had been killed by incubation in a boiling water bath (Fig.

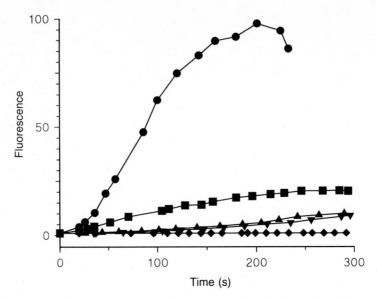

Fig. 8.4. Fluorescence of cells treated overnight with amphotericin B (0, ●; 0.01, ■; 0.1, ▲; and 1, ▼ μg/ml) and of boiled cells (◆) labelled with ChemChrome Y (1/5000 dilution). ChemChrome Y added at time 0. Fluorescence data normalized according to fluorescence of cells at time 0. The numbers of viable cells in the treated cell suspensions were 2.2×10^6, 9.6×10^5, 6.6×10^4, 0 and 0 cfu/ml respectively.

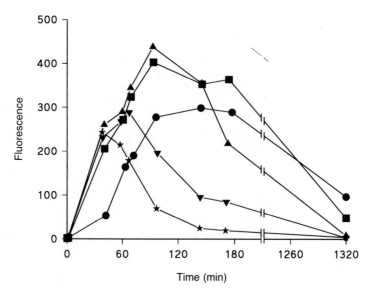

Fig. 8.5. Fluorescence of cells treated with amphotericin B and labelled with ChemChrome Y. Amphotericin B (0, ●; 0.1, ■; 1, ▲; 10, ▼; and 10, ★ μg/ml) and ChemChrome Y (1/5000 dilution) added to cell suspensions at time 0. Fluorescence data normalized according to fluorescence at time 0.

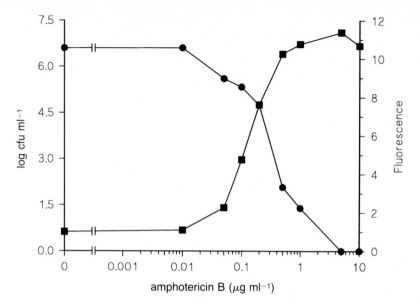

Fig. 8.6. Mean fluorescence of cells (■) and numbers of viable cells (●) in cell suspensions treated overnight with amphotericin B, then labelled with oxonol ($10\,\mu g$/ml) for 30 min. Fluorescence data normalized according to fluorescence of untreated cells.

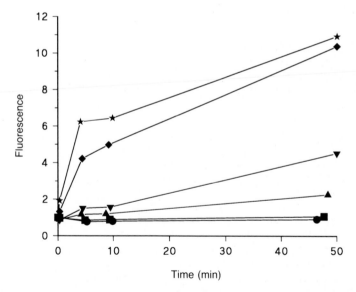

Fig. 8.7. Fluorescence of cells incubated with oxonol ($5\,\mu g$/ml) for 20 min then treated with amphotericin B. Amphotericin B (0, ●; 0.1, ■; 0.05, ▲; 0.1, ▼; 1, ◆; and 10, ★ μg/ml) added to cell suspensions at time 0. Fluorescence data normalized according to fluorescence at time 0.

8.4). Thus, the rate of ChemChrome Y cleavage appears to be a more sensitive marker of cell viability than propidium iodide exclusion. Conversely, when ChemChrome Y was added to cell suspensions at the same time as amphotericin B the treated cells initially became more fluorescent than untreated cells (Fig. 8.5). This suggests that:

1. Under the conditions of the test, there is a permeability barrier for ChemChrome Y at the level of the cell membrane which is breached by amphotericin B.
2. Uptake and/or cleavage of the dye is stimulated in the presence of amphotericin B.
3. There are changes within the cell caused by amphotericin B, such as in intracellular pH, which enhance fluorescence of the ChemChrome Y fluorochrome.

With increasing concentrations of amphotericin B the cells lost fluorescence after shorter intervals, suggesting leakage of the fluorochrome out of the cells, perhaps as a result of grosser changes in membrane permeability. Thus ChemChrome Y could be employed to monitor the short-term effects of amphotericin B on fungal cells, but may only be useful in this regard when the reasons underlying the higher fluorescence of the treated cells is fully understood.

Oxonol

Oxonol (used at 5 or $10 \mu g/ml$) stained untreated yeast cells lightly but stained yeast cells treated with amphotericin B much more brightly. The mean fluorescence of cells treated overnight with amphotericin B then labelled with oxonol increased with concentrations of amphotericin B of $0.05 \mu g/ml$ and above (Fig. 8.6). Whether this reflects a decrease in their

Table 8.2. MICs of amphotericin B

Conditions	MICs (μg/ml) of amphotericin B in:	
	Sabouraud's dextrose broth	Yeast nitrogen base
Amphotericin B alone	2	2
+ propidium iodide	4	2
+ oxonol	4	8
+ ChemChrome Y	2	1

MICs were determined in a conventional microtitre test, using 2-fold dilution series of amphotericin B, in two different media, in the presence and absence of fluorescent probes. Propidium iodide was added at $1 \mu g/ml$, oxonol at $10 \mu g/ml$ and ChemChrome Y at 1/5000 dilution. The inoculum was 10^5 cells/ml; plates were read by eye after overnight incubation at 37 °C.

membrane potential still has to be proved, but a decrease in membrane potential would be expected from the increases in membrane permeability caused by amphotericin B and as a natural consequence of cell death.

The fluorescence of cells changed quickly when amphotericin B at $1\,\mu g/ml$ and above was added to suspensions of cells to which oxonol had been added 20 min beforehand. Lower concentrations, 0.05 and $0.1\,\mu g/ml$, caused slower, but nonetheless obvious, increases in fluorescence (Fig. 8.7). Thus oxonol could be a sensitive probe of the more immediate effects of amphotericin B on cells, though whether the results are showing the development of a primary lesion or simply more general effects on cell viability is unclear.

None of the probes, at the concentrations used above, had much effect on MICs of amphotericin B (Table 8.2), or on the effect of amphotericin B on the growth of cultures in the biophotometer (data not shown).

Discussion

These results are promising but are really only a beginning. We have identified three probes – propidium iodide, ChemChrome Y and oxonol – that will be useful in investigating the interaction between amphotericin B and cells of *C. albicans*. ChemChrome Y and oxonol are the more sensitive. We now need to extend these studies to other strains, including resistant strains, other fungal species and other antifungal agents. We also need to work towards a better understanding of what the probes, especially oxonol, are telling us about the action of amphotericin B, but even at this early stage it is obvious that flow cytometry will be useful in the evaluation of antifungal agents.

It is unfortunate that the yeast cells did not stain with the ion-sensitive probes. It is possible that they lack the enzyme required to cleave the fluorogenic substrates to the ion-sensitive dyes; *Saccharomyces cerevisiae* is reported to lack the enzyme required to cleave Indo-1 but was labelled directly with the fluorescent product of Indo-1 using incubation at low pH (Halachmi and Eilam 1989). We intend to investigate the feasibility of this approach with *C. albicans*.

Others (Scott-Pore 1990; O'Gorman and Hopfer 1991) have looked at the use of flow cytometry for antifungal susceptibility testing of *C. albicans* with encouraging results (Scott-Pore used propidium iodide and Rose Bengal; O'Gorman and Hopfer used ethidium bromide). Like them, we noted changes in the scatter properties of organisms treated with amphotericin B, as well as in the uptake of selected probes. However, whereas effects on intrinsic properties of cells may vary with the drug, the probes are more likely to be generally applicable to work with all antifungal agents. Also, because the probes can be used with mammalian cells it should be possible to investigate the selectivity of action of amphotericin B in a mixed cell system – fungal cells mixed with mammalian cells.

Kukuruga et al. (1991) reported a flow cytometric method for screening isolates of *C. albicans* for the ability to form germ tubes and bind fibrinogen, both of which are implicated in pathology. It thus appears that flow cytometry could find many varied applications in mycology.

We have shown here that flow cytometry can be useful in studies of antifungal agents. At this stage the possibilities seem almost endless: rapid drug sensitivity testing of clinical isolates, investigation of modes of action and toxicity of antifungal agents including measuring selectivity of action between fungal and mammalian cells, study of efficacy of drug combinations, screening of novel compounds for specific antifungal activities, isolation of drug-resistant organisms from generally sensitive populations and their characterization, and so on. The future of the technique depends on its widespread acceptance and availability, standardization of test protocols, the development of the technique for use with filamentous fungi and, perhaps, the development of new probes specifically for use with fungi. Also, some work will be required to validate the use of probes, developed for work with mammalian cells, in studies with fungi. We believe, though, that flow cytometry will make a valuable addition to the battery of tests currently used to evaluate antifungal agents.

References

Brajtburg JK, Price HD, Medoff G, Schlessinger D, Kobayashi GS (1974) The molecular basis for the selective toxicity of amphotericin B for yeast and filipin for animal oils. Antimicrob Agents Chemother 5:377–382

Brajtburg JK, Powderley WG, Kobayashi GS, Medoff G (1990) Amphotericin B: current understanding of mechanisms of action. Antimicrob Agents Chemother 34:183–188

Butt TM, Hoch HC, Staples RC, St. Leger RJ (1989) Use of fluorochromes in the study of fungal cytology and differentiation. Exp Myco 13:303–320

Coleman K, Hunter PA, Ridgway Watt P (1983) A novel multichannel biophotometer and its use in determining antibiotic effects. In: Assessment of antimicrobial activity and resistance. Academic Press, London and New York

Gallis HA, Drew RH, Pickard WW (1990) Amphotericin B: 30 years of clinical experience. Rev Infect Dis 12:308–329

Halachmi D, Eilam Y (1989) Cystosolic and vacuolar Ca^{2+} concentrations in yeast measured with the Ca^{2+} sensitive fluorescence dye Indo-1. FEBS Lett 256:55–61

Jullien S, Capuozzo E, Salerno C, Crifo C (1991) Effects of polyene macrolides on the membrane potential of resting and activated human leukocytes. Biochem Int 24:307–319

Kukuruga M, Lynch M, Nakeff A, Sobel J (1991) Analysis of proliferation and differentiation of Candida albicans in vitro by multivariate flow cytometry. Cytometry Suppl. 5, abstr 41, p 34

O'Gorman MRG, Hopfer RL (1991) Amphotericin B susceptibility testing of Candida species by flow cytometry. Cytometry 12:743–747

Scott-Pore R (1990) Antibiotic susceptibility testing of Candida albicans by flow cytometry. Curr Microbiol 20:323–328

Shapiro HM (1988) Parameters and probes. In: Practical flow cytometry. Alan R Liss, New York, pp 115–198

Shapiro HM (1990) Flow cytometry in laboratory microbiology: new directions. ASM News Lett 584, November

Sokol-Anderson ML, Brajtburg J, Medoff G (1986) Amphotericin B-induced oxidative damage and killing of Candida albicans. J Infect Dis 154:76–83

Wilson HA, Chused T (1985) Lymphocyte membrane potential and Ca^{2+} sensitive potassium channels described by oxonol dye fluorescence measurements. J Cell Physiol 125:72–81

Chapter 9

Applications of Flow Cytometry in Bacterial Ecology

Clive Edwards, Julian P. Diaper, Jonathan Porter
and Roger Pickup

Introduction

Monitoring activity and enumeration of bacteria in natural environments has
always posed considerable problems. Conventional plating techniques have
normally been employed but have increasingly been shown to be of limited
value (Mills and Bell 1986) because only a small proportion of indigenous
species (*c.* 1%) can be isolated by any one technique (Jones 1977; Pickup
1991). The proposal that some bacteria can adopt a viable but non-culturable
state has further complicated the recovery of bacteria from natural environments (Roszak and Colwell 1987). In an attempt to circumvent these difficulties methods have been devised for the direct microscopic enumeration
of bacteria; for example, acridine orange staining which stains live bacteria
green while debris (and presumably dead cells) appears orange to red.
These methods, however, have a number of drawbacks that make them
unreliable (Postma and Altemuller 1990; Page and Burns 1991). The requirement for accurate and rapid methods for detecting target species present
within diverse and active populations has never been greater.

Concern about the effects of environmental pollution and climate change
are examples of physico-chemical constraints that may influence or shift
the natural balance of microbial species within natural ecosystems. Recent
debate has centred on the fate of genetically-engineered microorganisms
(GEMs) released into open environments. The impact of such potential
changes has yet to be fully realized, but concern centres on potential
modification of the microbial community structure resulting in elimination
(or stimulation) of pivotal species. Such an event has potentially disastrous
consequences in view of the role of key species in essential biogeochemical
cycles.

In this chapter we will describe the potential of flow cytometric methods
for enumerating viable bacteria, irrespective of whether they are culturable,
as well as methods for selectively identifying target species within mixed
populations. Other potential applications of flow cytometry will also be
discussed.

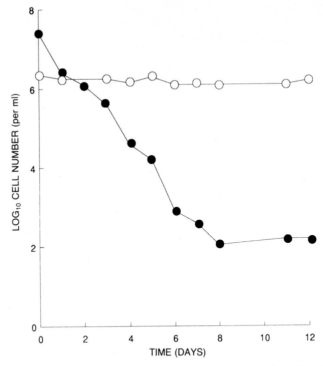

Fig. 9.1. Survival of *Aeromonas salmonicida* in filtered seawater monitored by measuring total counts using either flow cytometry (○) or colony forming units on nutrient agar plates (●).

Enumeration of Viable Bacteria

The problems associated with measurements of viable cells in natural environments are illustrated in Fig. 9.1. A fish pathogen, *Aeromonas salmonicida* was released into filtered seawater and incubated at 10 °C for 14 days. Samples were removed at intervals and cell numbers determined directly by flow cytometry or by counting colony forming units (cfu) after appropriate dilution and plating on to agar plates. Viable cells as measured by cfu fell rapidly during the first 8 days whereas total counts, as determined by flow cytometry, remained constant. There are two possible explanations for this result. The first is that cells die off but remain intact, hence the total count remains constant and the plate counts truly reflect the numbers of viable bacteria. This interpretation has tended to be the traditionally held one. The alternative explanation is that the disparity between the two methods of counting is due to dead intact cells together with a sub-population of viable but non-culturable cells the magnitude of which is unknown. This simple experiment illustrates very nicely the difficulty of drawing any meaningful conclusions from experiments that monitor bacterial numbers in natural environments. It is probable that properties of bacteria "in the wild" differ markedly from those in the laboratory. This conclusion is supported by the many recent studies of stimulus-sensor-regulator systems, of which the response of bacteria to starvation is a good example (see Matin 1990; and Gottschal 1990).

Membrane-Potential-Dependent Dyes

The staining of viable eukaryotic cells by fluorescent dyes taken up in response to membrane potential ($\Delta\Psi$) has been reported by a number of workers. The dyes include rhodamine 123 (Rh123; Darzynkiewicz et al. 1982) and 3,3'-dihexyloxacarbocyanine iodide (DiOC$_6$(3); Shapiro 1988). Rh123 has also been used to detect viable bacteria by microscopy (Bercovier et al. 1987; Matsuyama 1984). Fluorescein diacetate (FDA) is a lipophilic dye taken up by cells in which it is cleaved to fluorescein which is retained by cells that have an intact membrane. This dye has also been used as an indicator of viability (Shapiro 1988; Chrzanovski et al. 1984). We have assessed these three dyes for their suitability as indicators of viability by testing them against a range of bacteria and establishing the criteria by which flow cytometry may be used to confirm this. The rationale behind the tests is shown in Fig. 9.2.

Viable cells of *Staphylococcus aureus* were incubated with Rh123 in the presence or absence of chemicals that disrupted the functions of the cytoplasmic membrane thus making it leaky and unable to generate a $\Delta\Psi$. Only cells treated with the ionophore gramicidin are shown here and these exhibited a modal channel of fluorescence intensity of only 20. In the absence of the ionophore, *S. aureus* cells had a fluorescence distribution distinct from the gramicidin-treated culture with a modal channel of 100.

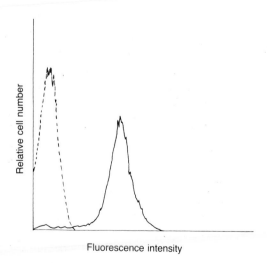

Fluorescence intensity

Fig. 9.2. Fluorescence histograms obtained by flow cytometry of *Staphylococcus aureus* stained with Rh123. The fluorescence distribution of cells treated with gramicidin S is represented by a *broken line*.

Other treatments, such as staining formaldehyde- or ethanol-killed cells, gave distributions very similar to that of gramicidin-treated cells (Fig. 9.2) showing that Rh123 cellular fluorescence depended upon generation of a membrane potential and therefore viability.

Table 9.1 summarizes the results obtained with a range of bacterial species with the three fluorescent dyes tested. Rh123 proved to be the most generally applicable and clearly demarcated fluorescence distributions were obtained for six of the bacterial species. *Bacillus subtilis* and *Aeromonas salmonicida* could not be stained at all by Rh123 whilst *Pseudomonas fluorescens* was only partially stained; some 70%–80% of *Aeromonas hydrophila* cells could routinely be stained. $DiOC_6(3)$ was taken up by all the species except for *Aeromonas salmonicida*. However, treatment with ionophores or fixatives could only partially resolve non-viable cells, implying a high degree of non-specific binding to cell structures by this dye. In the extreme case of *A. hydrophila*, cells exhibited the same amount of fluorescence with or without ionophores or other treatments that killed the cells. Prior to treatment with Rh123 and $DiOC_6(3)$, gram-negative species were washed in EDTA-containing buffers in order to disrupt the outer membrane and facilitate entry of the dye. No reduction in viability occurred with EDTA treatment.

Fluorescein-stained bacteria could not be detected by flow cytometry after incubation in the presence of FDA; the exception was *B. subtilis*, for which this dye was shown to be an excellent indicator of viability. FDA normally enters the cell and is cleaved by esterase activity which releases the fluorescein. Analysis of the species tested showed that, apart from *Escherichia coli*, all possessed an esterase activity capable of cleaving FDA. Presumably, except in *B. subtilis*, fluorescein could not be retained by the cells long enough for them to be analysed by flow cytometry. Leakage of fluorescein from cells in this way has also been reported by Lundgren (1981).

The results shown in Table 9.1 indicate that, even in a relatively small number of bacterial species, there is no universal viability stain. This no

Table 9.1. Measurement of bacterial viability by fluorescent dyes

	Rhodamine 123 (Rh123)	Dihexyloxacarbocyanine ($DiOC_6(3)$)	Fluorescein diacetate[a] (FDA)
B. subtilis	−	+/−[b]	+
S. aureus	+	+/−	−
E. coli	+	+/−	−
Salmonella sp.	+	+/−	−
A. hydrophila	+ (70%–80%)	−, but cells stained	−
A. salmonicida	−	−	−
P. fluorescens	10% cells labelled	+/−	−
E. herbicola	+	+/−	−
A. globiformis	+	+/−	−
A. vinelandii	+	+/−	−

[a] All bacteria *except E. coli* have esterase activity, but none apart from *B. subtilis* retain the dye.
[b] +/− indicates that fluorescent live cells could not be clearly distinguished from dead ones.

doubt reflects the differences in cell surface physiology of bacteria and would explain the failure of all three dyes to fully stain the aeromonads, which are known to possess unusual surface layers such as the A-protein of *A. salmonicida* (Adams et al. 1988). Pinpointing viability of a single species within a heterogeneous population would require a fluorescent signature. A possible candidate would be the use of fluorescent antibody probes and a model system developed to show their applicability is described below.

Enumeration of Immunofluorescently Labelled Bacteria

Molecular-based methods are of much current interest for detection of bacteria in natural environments. Currently, the limits of detection of such methods as luminometry (*lux* gene), DNA hybridization, polymerase chain reaction or fluorescent oligonucleotides is no better than 10^3 bacteria per millilitre water or per gram soil. However, fluorescent antibodies exhibit much more sensitive lower limits of around 2–10 cells per millilitre water or per gram soil (Pickup 1991). In this section we describe the use of flow cytometry to detect and count *Staphylococcus aureus* labelled with FITC-IgG. *S. aureus* is an important pathogen that cannot be quickly identified by coventional plating techniques (Harvey and Gilmour 1985). Protein A is a cell wall immunoglobulin-binding protein expressed by 98% of *S. aureus* strains (Harlow and Lane 1988). The ability of this protein to bind fluorescently-labelled (FITC) human IgG was used as a model system to assess flow cytometry for detection and enumeration of *S. aureus*.

Fig. 9.3a shows *S. aureus* cells stained with FITC-IgG and detected by flow cytometry. Controls whereby FITC-IgG was tested against a range of other bacteria showed no non-specific binding; *Pseudomonas fluorescens* reacted in this way as shown in Fig. 9.3b. Fig. 9.3c shows cells of *S. aureus* released into eutrophic lakewater that contained approximately 10^6 to 10^7 bacteria/ml and detected flow cytometrically after incubation with FITC-IgG. The counts of *S. aureus* calculated from this fluorescence distribution were in close agreement with those obtained from plate counts, showing the high degree of specificity of the method. Further experiments revealed that the lower limit of detection in lakewater was around 10^3 *S. aureus* cells/ml. This could be greatly improved by concentration of water samples by filtration or centrifugation. In pure cultures the lower limit was approximately 10^1 to 10^2 cells/ml.

Other Applications

Table 9.2 lists other potential applications of flow cytometry to studies of bacterial ecology. Most of these methods have already been applied to eukaryotic systems and a few to some bacterial species in pure culture. Flow cytometry has been used to study the bacterial cell cycle with respect to

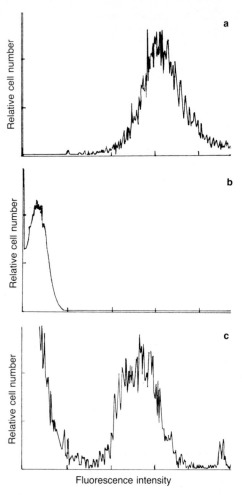

Fig. 9.3a–c. Flow cytometric determination of fluorescent labelling of bacteria with FITC-IgG. **a** *Staphylococcus aureus*; **b** *Pseudomonas fluorescens*; **c** *S. aureus* in eutrophic lakewater.

DNA replication, chromosomal number and the effects of antimicrobial agents on growth and division (Steen et al. 1982, 1990). Measurements of bacterial chromosome number per unit cell give an indication of the proportion of a population that is growing or senescent. Hoechst 33342 binds to AT-rich regions of DNA and is therefore a specific dye for DNA, and should allow such measurements to be made of bacteria in natural environments. Additional information regarding metabolic status of bacteria is possible using pyronin Y for total RNA. RNA levels respond rapidly to changing growth rates of bacteria and during starvation decrease rapidly. Similar information is possible using fluorescein isothiocyanate (FITC) as a stain for total protein. However, in our hands the information provided by both RNA and total protein staining only gives information of a very general and qualitative nature; flow cytometric measurements of these

Table 9.2. Potential applications of flow cytometry in bacterial ecology

Application	Fluorescent parameter	Target
Cell cycle events	Hoechst 33342	AT regions of DNA
Chromosomal number	Mithramycin ethidium bromide	DNA
Metabolic status of cells	Pyronin Y	RNA content
	FITC	Total protein
	Autofluorescence	Flavin and pyridine nucleotides
Methanogens	Autofluorescence	F_{420} (deazaflavin)
Cyanobacteria	Autofluorescence	Photosynthetic pigments
Species-specific identification	FITC-conjugated oligonucleotides	DNA, 16S rRNA
	DIG-FITC-conjugated antibody	16S rRNA
	FITC- or phycoerythrin-conjugated polyclonal/monoclonal antibodies	Antigens (surface, flagellar)

macromolecules should be underpinned by chemical assays. Autofluorescence of cells excited at appropriate wavelengths is probably due to flavin and pyridine nucleotides. Such analysis coupled to light scatter data can be used to analyse changing redox states within a bacterial population. In environmental samples such studies could be linked to tagging a specific organism of interest by the more discriminating antibody or nucleic acid probes. Methanogens give a sufficiently strong autofluorescence as to be easily detected by flow cytometry. This may provide a rapid and easy approach for enumerating their numbers in soil profiles. Such an approach could also be used for enumeration of photosynthetic bacteria from natural environments by using autofluorescence of their pigments. Similar contributions from unicellular algae may be gated out on the basis of their greater light scatter.

Probably one of the most exciting new applications of flow cytometry is the detection and identification of bacteria labelled with highly specific fluorescent molecular probes. These principally include antibodies and oligonucleotides. Nucleic acid probes have become a powerful means for detecting microorganisms and for examining their interactions in ecosystems (Saylor and Layton 1990; Stahl et al. 1988). Specific identification of individual microbial cells by flow cytometry has been reported using fluorescent oligonucleotide probes against intracellular 16S rRNA (Amann et al. 1990). However, sensitivity of this approach is limited at present due to the low fluorescent signals obtained from cells isolated from natural environments. This probably reflects the low rRNA copy number due to non-growing starvation conditions and the fact that only one dye molecule can be bound per probe. Recently, Zarda et al. (1991) have attempted to improve the amount of fluorescence by labelling oligonucleotides with digoxigenin (DIG) which after hybridization may be detected using fluorescently labelled antibodies. Since the binding proteins may be labelled with several fluorescent molecules, an increase in sensitivity should be detected. However, early work in this area using flow cytometry has not been very promising and little to no enhancement of fluorescence was obtained using three test bacterial species (Zarda et al. 1991). Applications of antibody probes have already been discussed and potentially they may herald a more sensitive means of detecting low numbers of target bacteria within natural environments. The work reported here for *S. aureus* supports this as does recent work by Page and Burns (1991), who demonstrated the applicability of flow cytometry for detecting soil bacteria using FITC-labelled secondary antibodies against monoclonal antibodies as the primary recognition agents.

Conclusions

Developments in flow cytometric analyses for bacteria are in their infancy. This is because the instruments and techniques were developed primarily for eukaryotic cells. Here we have attempted to demonstrate, using model systems, how flow cytometry may be used to answer fundamental questions concerning not only bacterial ecology but other areas such as microbial physiology, pathogenicity and biotechnology. As techniques become more refined, no doubt other applications will be found. It seems to us that the developing methodologies in contrast to observations made under more artificial and conventional laboratory conditions, could revolutionize our understanding of how bacteria behave in natural environments.

Acknowledgement

This work was supported by grants from the Natural Environment Research Council.

References

Adams A, Bundy A, Thompson K, Horne MT (1988) The association between virulence and cell surface characteristics of *Aeromonas salmonicida*. Aquaculture 69:1–14

Amann RI, Binder BJ, Olson RJ, Chisolm SW, Devereux R, Stahl DA (1990) Combination of 16S rRNA-targeted oligonucleotide probes with flow cytometry for analysing mixed microbial populations. Appl Environ Microbiol 56:1919–1925

Bercovier H, Resnick M, Kornitzer D, Levy L (1987) Rapid method for testing drug-susceptibility of *Mycobacteria* spp. and Gram-positive bacteria using Rhodamine 123 and fluorescein diacetate. J Microbiol Meth 7:139–142

Chrzanowski TH, Crotty RD, Hubbard JG, Welch RP (1984) Applicability of the fluorescein diacetate method of detecting active bacteria in freshwater. Microbial Ecol 10:179–185

Darzynkiewicz Z, Traganos F, Stainio-Coico L, Kapuscinski J, Melamed MR (1982) Interactions of Rhodamine 123 with living cells studied by flow cytometry. Cancer Res 42:799–806

Gottschal JC (1990) Phenotypic response to environmental changes. FEMS Microbiol Ecol 74:93–102

Harlow E, Lane D (1988) Antibodies: a laboratory manual. Cold Spring Harbor Laboratory, New York

Harvey J, Gilmour A (1985) Application of current methods for isolation and identification of staphylococci in raw bovine milk. J Appl Bacteriol 59:207–221

Jones JG (1977) The effect of environmental factors on estimated viable and total populations of planktonic bacteria in lakes and experimental enclosures. Freshwater Biol 7:61–97

Lundgren B (1981) Fluorescein diacetate as a stain of metabolically active bacteria in soil. Oikos 36:17–22

Matin A (1990) Molecular analysis of starvation stress in *Escherichia coli*. FEMS Microbiol Ecol 74:185–196

Matsuyama T (1984) Staining of bacteria with Rhodamine 123. FEMS Microbiol Lett 21:153–157

Mills AL, Bell PE (1986) Determination of individual organisms and their activities in situ. In: Tate RL (ed) Microbial autoecology: a method for environmental studies. Wiley, New York, pp 27–60

Page S, Burns RG (1991) Flow cytometry as a means of enumerating bacteria introduced into soil. Soil Biol Biochem 23:1025–1028

Pickup RW (1991) Development of methods for the detection of specific bacteria in the environment. J Gen Microbiol 137:1009–1019

Postma J, Altemuller HJ (1990) Bacteria in thin soil sections stained with the fluorescent brightener calcofluor white M2R. Soil Biol Biochem 22:89–96

Roszak DB, Colwell RR (1987) Survival strategies of bacteria in the natural environment. Microbiol Rev 51:365–379

Saylor GS, Layton AC (1990) Environmental application of nucleic acid hybridization. Annu Rev Microbiol 44:625–648

Shapiro HM (1988) Practical flow cytometry, 2nd edn. Alan R Liss, New York

Stahl DA, Flesher B, Mansfield HR, Montgomery L (1988) Use of phylogenetically based hybridization probes for studies of ruminal microbial ecology. Appl Environ Microbiol 54:1079–1084   ·

Steen HB, Skarstad K, Boye E (1990) DNA measurements of bacteria. Meth Cell Biol 33:519–526

Steen HB, Boye E, Skarstad K, Bloum B, Godal T, Mustafa S (1982) Applications of flow cytometry on bacteria: cell cycle kinetics, drug effects and quantitation of antibody-binding. Cytometry 2:249–257

Zarda B, Amman R, Wallner G, Schleifer K-H (1991) Identification of single bacterial cells using digoxigenin-labelled rRNA-targeted oligonucleotides. J Gen Microbiol 137:2823–2830

Chapter 10

Analysis of Microalgae and Cyanobacteria by Flow Cytometry

Alex Cunningham

Introduction

Flow cytometers first became available commercially around 1970, and early studies of algal cultures were carried out by Paau et al. (1978), Price et al. (1978) and Trask et al. (1982). A workshop report by Yentsch et al. (1983b) drew attention to the remarkable power of flow cytometry for aquatic particle analysis, and this was followed by the rapid adoption of the technique in oceanographic research. A special volume of *Cytometry* (Yentsch and Horan 1989) has been devoted to this topic, and the proceedings of a lively NATO Advanced Study Institute have recently been published (Demers 1992).

In contrast, the use of flow cytometry for freshwater phytoplankton research has been confined so far to a few European laboratories. This may be due to the fact that phytoplankton assemblages vary significantly in their amenability to flow cytometric analysis. Oceanic forms are often single-celled, rarely more than a few tens of micrometres in diameter, and are easily analysed by standard flow cytometers. On the other hand, a freshwater sample may contain cells and colonies spanning a range of around two orders of magnitude in diameter and three orders of magnitude in length (Peeters et al. 1989). Cytometric equipment that has been designed for mammalian cells, particularly blood cells, is not ideally suited to the analysis of particles with such a wide range of characteristics.

Reviews of the application of flow cytometry to microalgal research have been published by Cunningham and Leftley (1986), Burkill (1987), Burkill and Mantoura (1990) and Yentsch (1990). There is clearly no need to repeat this material, and the present chapter is written with other objectives in mind. One is to suggest that flow cytometry has the potential to make further major contributions to the study of microalgal physiology and ecology. Another is to point out that the term flow cytometer covers a range of instruments with quite different analytical capabilities: the widespread adoption of flow cytometry in microalgal research would be facilitated by a clearer understanding of the way in which instrument and cell interact to determine the information content of cytometric data sets.

Microalgae and Cyanobacteria as Cytometric Objects
Size and Shape

Photosynthetic planktonic organisms exhibit a great range of cell sizes, from $1-2\,\mu m$ for picoplankton and prochlorophytes to $80-100\,\mu m$ for large freshwater species. The range of cell shapes is spectacular, from simple rods and ellipsoids to elaborate asymmetrical structures: genera such as *Staurastrum* and *Ceratium* provide objects that have no counterpart in biomedical cytometry. Furthermore many species form chains, filaments, semi-regular colonies or globular masses the overall dimensions of which may run to hundreds of micrometres. Fig. 10.1 shows an outline sketch of some common freshwater phytoplankton, with the beam height of a typical laser-illuminated flow cytometer drawn in for comparison. The optical signals measured will be influenced not only by the relative dimensions of beam and particle, but also by the orientation of the larger objects in the beam. Individual phytoplankton cells are fairly rigid and not easily deformed in flow cells, but colonies vary in their degree of adhesion, and the looser aggregates may fragment during sample injection or analysis.

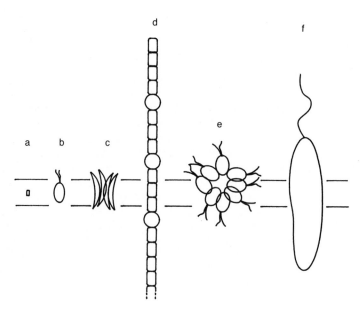

Fig. 10.1a–f. Typical freshwater phytoplankton sketched roughly to scale, with horizontal lines indicating the typical beam waist height $(10\,\mu m)$ in a laser-illuminated flow cytometer. Genera shown: **a** *Synechococcus*; **b** *Chlamydomonas*; **c** *Scenedesmus*; **d** *Anabaena*; **e** *Eudorina*; **f** *Euglena*.

Autofluorescence from Pigment Systems

All phytoplankton cells contain chlorophyll a, the primary photosynthetic pigment, which has absorption bands in the blue (400–450 nm) and in the red (at 650–700 nm) and fluoresces in the far red (roughly 680–720 nm). Most cells also contain accessory pigments that capture photons in the middle of the visible spectrum and pass the energy to chlorophyll a. The nature of the accessory pigments varies from one algal group to another: they include chlorophyll b (green algae), chlorophyll c and carotenoids (diatoms and dinoflagellates) and phycobiliproteins (cryptophytes and cyanobacteria). A comprehensive review of absorption and fluorescence spectra for both individual photopigments and intact cells may be found in Prezelin and Boczar (1986). One of the phycobiliproteins, phycoerythrin, fluoresces strongly at around 580 nm, and this provides a useful diagnostic feature for cyanobacteria and cryptophytes (Yentsch and Yentsch 1979), provided that they are in a suitable state of photoadaptation (Hilton et al. 1988). Other accessory pigments can be detected in flow cytometry only by their influence on the excitation spectrum for chlorophyll a fluorescence.

One advantage of measuring the fluorescence of photopigments in a flow cytometer is that cells are exposed to very short periods of illumination (of the order of 10 μs) and photobleaching is minimized. This facilitated the discovery of widespread populations of oceanic prochlorophytes that were difficult to detect by fluorescence microscopy (Chisholm et al. 1988; Li and Wood 1988).

Light Scattering

Most phytoplankton cells are one or two orders of magnitude larger than the wavelength of visible light (0.4–0.7 μm). When absorbing particles in this size range are exposed to collimated illumination, the most significant scattering mechanism is diffraction in a narrow forward lobe, within 2° or 3° of the axis of illumination (van de Hulst 1957). Scattering at wider angles may occur from spines and other fine morphological details, and from internal structures such as gas vacuoles (Dubelaar et al. 1987). Fig. 10.2 shows forward scattering measurements for a *Dunaliella* cell mounted statically in a laser beam of relatively low divergence: the magnitude of the forward scattering signal is obviously very sensitive to the size of beam stop used in a flow cytometer. The azimuthal distribution of intensity in the narrow-angle forward scattering pattern can be used as an indicator of particle morphology (Cunningham and Buonnacorsi 1992). Depolarization of forward-scattered laser illumination by single cells varies from one algal class to another, and is particularly marked for coccolithophorids (Olson et al. 1989).

Fluorochrome Staining

There were early fears that standard fluorescent stains would be difficult to use on algal cells because of thick cell walls and strong autofluorescence, but

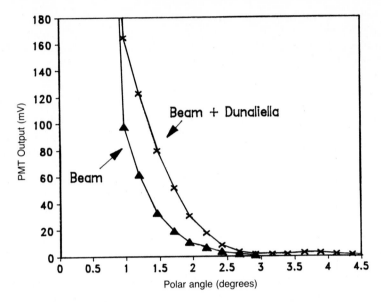

Fig. 10.2. Measured light intensities in the forward direction for a laser beam brought to a 200 μm diameter focus. Most of the forward scattered light from a 9.1 μm diameter *Dunaliella* cell falls within the divergence angle of the laser beam, and in practice would be obscured by a beam stop.

these have generally proved to be unfounded. Potential interference from photopigment fluorescence can be avoided by fixing cells in ethanol (which extracts pigments) or by photo-oxidation (Yentsch et al. 1983a), while cell permeability can be increased using non-ionic detergents. Quantitative DNA staining in microalgae has been carried out using propidium iodide (Hutter and Eipel 1979; Vaulot et al. 1986), ethidium bromide (Paau et al. 1978), mithramycin (Yentsch et al. 1983a) or Hoechst 33258 (Bonaly et al. 1987). Protein can be stained using fluorescein isothiocyanate (Paau et al. 1978; Hutter and Eipel 1979), and the cell content of neutral lipids can be quantitatively determined using nile red (Cooksey et al. 1987). Cell viability, and possibly the degree of metabolic activity, can be estimated from the rate at which cells cleave fluorescein diacetate (Dorsey et al. 1989). Toxin production in at least one diatom can be determined by flow cytometry following oxidation of saxitoxin to a fluorescent derivative (Yentsch 1981).

Instrumentation for Microalgal Studies

Comprehensive discussions of the principles of flow cytometer construction can be found in van Dilla et al. (1985) and Shapiro (1988), while Darzynkiewicz and Crissman (1990) and Ormerod (1990) give up-to-date reviews of analytical methods and staining protocols. There are two major differences between the requirements of microalgal analysis and the features found on standard biomedical machines: the range of signals and particle sizes is much larger for the former, and the wavelengths that are optimal for stimulating photopigment autofluorescence are not the same as those normally used for exciting cytochemical stains.

Light Sources and Optics

Light sources for flow cytometry can be characterized by their intensity, wavelength, and degree of collimation. Much of the early work on the flow cytometric analysis of algal cells was carried out using instruments illuminated by multiwatt water-cooled argon lasers, and such instruments are still taken to sea (see Chapter 11). However, there is now ample evidence that air-cooled lasers producing 10–50 mW of blue light are sufficiently powerful for virtually all microalgal work, including the detection of picoplankton and prochlorophytes (Cunningham 1990; Olson et al. 1990; Li et al. 1991). High-pressure mercury arc lamps are a practical alternative to air-cooled lasers, and instruments based on these have been used successfully for studies on both larger algal cells and picoplankton (Olson 1983; Li and Wood 1988; Børsheim et al. 1989). It must be noted, however, that instrument sensitivity is not solely determined by the power of the light source: it is also a function of the efficiency of the beam delivery and fluorescence collection optics. Differences in the numerical aperture of fluorescence collecting lenses, and in the presence and size of obscuration bars, can lead to great differences in sensitivity between commercial instruments using similar light sources.

Lasers are essentially monochromatic, and therefore are efficient light sources for flow cytometry only if the wavelength at which they operate matches the absorption band of a fluorochrome fairly closely. In cytochemistry, a great deal of attention has been paid to this problem, and most modern dyes are excitable by the main argon lines at 488 and 514 nm. It is possible to excite chlorophyll fluorescence directly at 488 nm, but the optimum wavelength is much shorter and the helium–cadmium line at 441 nm would be a better choice. Chlorophyll fluorescence can also be excited through light absorption by accessory pigments: for example peridinin and fucoxanthin at 514 nm (argon ion) and phycocyanins at 633 nm (helium–neon). Mercury arc lamps have strong emission lines at 405 and 436 nm, and are relatively efficient exciters of chlorophyll fluorescence.

Lasers produce highly collimated beams with a Gaussian distribution of intensity. They can be brought to a tight focus and are well suited for exciting small cells and for making scattering measurements, but they demand great precision in the placing of the cells in the beam waist and large cells are usually only partly illuminated at a given instant. Arc lamps, on the other hand, produce highly divergent light, and when employed in Koehler illumination systems produce even illumination of the field of view. The light power delivered to the sample analysis zone from an arc lamp may exceed that from a small laser, but the range of wavelengths and the area illuminated are much greater. One advantage of Koehler illumination is that the dimensional tolerance in particle placement is relatively high, and large cells can be evenly illuminated. However, light scattering measurements in arc lamp systems require complicated optics (Steen 1986), and are less easily interpreted in physical terms than those from a monochromatic laser beam (Buonaccorsi and Cunningham 1990).

The use of multiple, spatially separated excitation wavelengths allows the discrimination of major algal groups by their different accessory pigment contents (Hilton et al. 1989; Hofstraat et al. 1990). This technique is usually associated with large flow cytometers equipped with two or sometimes three lasers, but Heiden et al. (1990) have shown that it is technically feasible with a single arc lamp.

Flow Cells and Sample Handling

A flow cell for phytoplankton analysis should resist clogging by the range of particle sizes normally encountered. It should not break up colonies and filaments, and elongated objects should be aligned with their long axis parallel to the direction of flow. Planar optical windows increase the sensitivity of fluorescence measurements (by avoiding the need for obscuration bars to intercept refracted laser light) and allow more accurate scattering measurements. These design objectives are all achievable in purpose-built flow cells with wide orifices or channels for hydrodynamic focussing and low longitudinal velocity gradients in the fluidic system (Peeters et al. 1989; Cunningham 1990). On the other hand, high-accuracy DNA measurements and precise cell positioning require narrow sample-bearing cores. These are usually produced by small orifices and high velocity gradients, and solutions to this design incompatibility are currently being sought. There is little published work on the design of special flow cells for phytoplankton analysis in arc lamp instruments, but some existing flow cells permit electrical volume measurements to be made at the same time as optical measurements: this feature is of immense value in aquatic particle analysis.

Much microalgal research requires absolute particle concentrations to be measured. This is easily achieved by injecting the sample via a motorized syringe or by sucking a known volume through a gear pump, but some commercial instruments do not offer these features. Since many algal cells sink or float in the sample holding chamber, gentle stirring may be required.

Data Acquisition

The wide dynamic range of optical signals from algal cells means that logarithmic amplifiers are required for virtually all work except laboratory studies of pure cultures. In laser instruments where the beam is brought to a tight focus in one dimension, larger cells and colonies are effectively scanned by an illuminating strip. The shape of the pulses obtained under these conditions carries a great deal of information on particle morphology (Cunningham 1991) that is not available from current designs of arc-lamp instruments. However, if signals relating to whole particles are required then the fluorescence pulses must be integrated electronically.

In analysing phytoplankton samples from coastal or inland waters, it is possible that the dynamic range of the integrated signals will exceed the capabilities of standard electronics in commercial instruments: an interesting hybrid (digital/analogue) solution to this problem is described by Dubelaar et al. (1989).

Software

A research-grade flow cytometer may measure between 5 and 10 parameters per particle. However, most of the analytical protocols developed for biomedical purposes use only one or two optical parameters (Darzynkiewicz and Crissman 1990; Ormerod 1990), and in such cases data analysis can be carried out as sampling proceeds. In contrast, the analysis of natural phytoplankton samples is essentially multiparametric (Hofstraat et al. 1990) and most workers collect the signals in list mode and analyse them later. The software packages available from instrument manufacturers and independent suppliers for analysing multiparametric data sets require active operator involvement in setting thresholds and identifying clusters, and the time for data analysis often greatly exceeds that required to pass the samples through the flow cytometer. One way round this software block might be the application of simulated neural networks (Frankel et al. 1989; Balfoort et al. 1992). However, these require extensive training on data sets from pure cultures, and improvements in conventional statistical treatments are also required.

Size and Portability

The major barriers to the extensive field use of flow cytometers are the physical size and power requirements of existing instruments. A significant reduction in size and weight has recently been achieved by the use of switched-mode power supplies for argon lasers and arc lamps, and further reductions can be expected as manufacturers incorporate portable computers and miniaturized electronics into their new instruments. Flow cytometers equipped with single air-cooled lasers require at least 1 kW of electrical power, while the lower limit for an arc-lamp instrument is probably around 250 W. A rather unexpected design problem arises from the fact that helium–neon and helium–cadmium lasers are much longer than argon lasers of similar power, and do not lend themselves to compact instrument construction.

Design and Function

The signals obtained from a flow cytometer are not simple functions of the optical characteristics of cells or colonies. They are heavily influenced by the excitation and collection geometry of the instrument, by the wavelengths and filters used in the analysis, and by the electronics used for signal conditioning and data acquisition. Fig. 10.3 shows how the ability of a flow cytometer to distinguish between two morphologically dissimilar species on the basis of chlorophyll fluorescence depends critically on the electonic processing applied to the photomultiplier output. As a result, the infor-

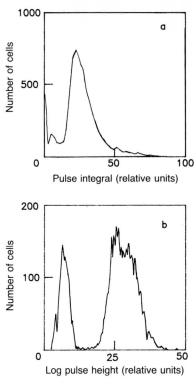

Fig. 10.3a,b. Red fluorescence signals (wavelength >650 nm) from a mixture of *Anabaena solitaria* and *Scenedesmus quadriculata* cells. The two species cannot be distinguished by integrated pulse measurements (**a**), but pulse height measurements are clearly bimodal (**b**).

mation obtained can be surprisingly instrument-specific. Some parameters that have proved to be valuable in algal research, such as beam transit time, pulse height/integral ratio, Coulter volume, beam depolarization, and azimuthally resolved forward scattering, may not measurable on unmodified commercial instruments.

Areas of Application

Flow cytometry is already established as an important tool in aquatic ecology. At the simplest level, the use of chlorophyll fluorescence as a trigger for data acquisition makes it is possible to count and analyse algal cells in the presence of large quantities of detrital particles: this is often taken for granted, but it represents a major breakthrough in enumerating algal cells in turbid waters. The rapidity with which large numbers of samples can be analysed makes the technique very suitable for the temporal and spatial profiling of natural populations, and for environmental

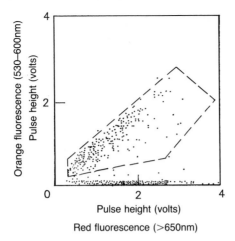

Fig. 10.4. Bivariate analysis of a nitrogen-starved culture of the marine diatom *Phaeodactylum tricornutum* stained with nile red and excited at 488 nm. Data for 1024 cells are shown: 32% of the population (falling within the pentagonal area) exhibited significant yellow/orange fluorescence, and this corresponded to the proportion found to contain lipid droplets on microscopical examination.

monitoring. Picoplanktonic and filamentous cyanobacteria, prochlorophytes and coccolithophorids have distinctive cytometric signatures, and finer discriminations (down to genus level) are sometimes possible. This means that flow cytometry can be used to investigate species succession and competition, and also for studies of selective grazing by protozoans, crustaceans and molluscs (Cucci et al. 1985, 1989; Gerritsen et al. 1987).

Flow cytometry may be expected to play an increasing role in ecotoxicology: toxins such as trace metals and pesticides inhibit algal fluorescence, and this may provide a more rapid assay of toxicity than current procedures that measure growth rates over several days in culture (Hutter et al. 1980; Samson and Popovic 1988). The technique has also been used to detect pollutant stress on symbiotic algal cells in lichens (Berglund and Eversman 1988).

In the area of cell physiology, flow cytometry has already made significant contributions to studies of microalgal and cyanobacterial cell cycle dynamics (Chisholm et al. 1986; Lefort-Tran et al. 1987). However, the diverse life histories exhibited by these organisms, which include sexual reproduction and spore formation, suggest that this is a fertile field for further research. Improvements in cytochemical staining, and the production of fluorescein-conjugated antibodies to surface markers and intracellular enzymes, promise to extend the range of flow cytometric techniques (Campbell and Carpenter 1987; Yentsch 1990). Measurements of the accumulation of intracellular products by flow cytometry may find applications for process control and strain selection in biotechnology: for example, Fig. 10.4 shows the variation in lipid droplet synthesis in a culture of the marine diatom *Phaeodactylum tricornutum*.

Conclusions

Algal and cyanobacterial cells are well suited to analysis by flow cytometry, and the instruments could be employed with advantage in any aquatic sciences laboratory that currently uses a Coulter counter or a fluorescence microscope. Widespread adoption of the technique may currently be hindered by perceptions of the high cost, large size and cumbersome operating procedures of first-generation instruments, but the technology has advanced markedly in recent years. The rapidly growing literature on the subject shows that a great deal of microalgal and cyanobacterial research can be tackled with standard or slightly customized commercial equipment. Nevertheless, there is a requirement for instruments designed specifically for phytoplankton analysis: features currently under development include quantitative scattering measurements, specialized pattern recognition software, the utilization of solid-state light sources, and particle imaging in flow. With these innovative techniques, flow cytometry will continue to make substantial contributions to our understanding of the biology and ecology of photosynthetic microorganisms.

Acknowledgements

The author has benefited greatly from discussions on algal physiology with J.W. Leftley (Dunstaffnage Marine Laboratory) and on the cytometric analysis of freshwater phytoplankton with H.W. Balfoort (University of Amsterdam).

References

Balfoort HW, Snoek J, Smits JRM, Breedveld LW, Hofstraat JW, Ringelberg J (1992) Automatic identification of algae: neural network analysis of flow cytometric data. J Plankton Res (in press)

Berglund D, Eversman S (1988) Flow cytometric measurement of pollutant stresses on algal cells. Cytometry 9:150–155

Bonaly J, Bre MH, Lefort-Tran M, Mestre JC (1987) A flow cytometric study of DNA staining in situ in exponentially growing and stationary *Euglena gracilis*. Cytometry 8:42–45

Buonnacorsi GA, Cunningham A (1990) Azimuthal inhomogeneity in the forward light scattered from microalgal colonies, and its use as a morphological indicator in flow cytometry. Limnol Oceanogr 35:1170–1175

Burkill PH (1987) Analytical flow cytometry and its application to marine microbial ecology. In: Sleigh MA (ed) Microbes in the sea. Ellis Horwood, Chichester, pp 139–166

Burkill PH, Mantoura RFC (1990) The rapid analysis of single marine cells by flow cytometry. Phil Trans R Soc Lond A 333:49–61

Børsheim KY, Harboe T, Johnsen T, Norland S, Nygaard K (1989) Flow cytometric characterisation and enumeration of *Chrysochromulina polylepis* during a bloom along the Norwegian coast. Mar Ecol Prog Ser 54:307–309

Campbell L, Carpenter EJ (1987) Characterization of phycoerythrin-containing *Synechococcus* spp. populations by immunofluorescence. J Plankton Res 9:1167–1181

Chisholm SW, Olson RJ, Zettler ER, Goericke R, Waterbury JB, Welschmeyer NA (1988) A novel free-living prochlorophyte abundant in the oceanic euphotic zone. Nature 334:340–343

Chisholm SW, Armbrust EV, Olson RJ (1986) The individual cell in phytoplankton ecology: cell cycles and applications of flow cytometry. Can Bull Fish Aquat Sci 214:343–369

Cooksey KE, Guckert JB, Williams SA, Callis PR (1987) Fluorimetric determination of the neutral lipid content of microalgal cells using nile red. J Microbiol Meth 6:333–345

Cucci TL, Shumway SE, Newell RC, Selvin R, Guillard RL, Yentsch CM (1985) Flow cytometry: a new method for characterization of differential ingestion, digestion and egestion by suspension feeders. Mar Ecol Prog Ser 24:201–204

Cucci TL, Shumway SE, Brown WS, Newell CR (1989) Using phytoplankton and flow cytometry to analyze grazing by marine organisms. Cytometry 10:658–669

Cunningham A (1990) A low cost, portable flow cytometer specifically designed for phytoplankton analysis. J Plankton Res 12:149–160

Cunningham A (1991) Fluorescence pulse shape as a morphological indicator in the analysis of colonial microalgae by flow cytometry. J Microbiol Meth 11:27–36

Cunningham A, Buonnacorsi GA (1992) Narrow-angle forward light scattering from individual algal cells: implications for size and shape discrimination in flow cytometry. J Plankton Res 14:223–234

Cunningham A, Leftley JW (1986) Application of flow cytometry to algal physiology and phytoplankton ecology. FEMS Microbiol Rev 32:159–164

Darzynkiewicz Z, Crissman HA (1990) Flow cytometry. Methods in cell biology 33. Academic Press, New York

Demers S (ed) (1991) Particle analysis in oceanography: proceedings of a NATO Advanced Study Institute at Acquefreda di Maratea, Italy, November 1990. Springer, Berlin Heidelberg New York

Dorsey J, Yentsch CM, Mayo S, McKenna C (1989) Rapid analytical technique for the assessment of cell metabolic activity in marine microalgae. Cytometry 10:622–628

Dubelaar GBJ, Visser JWM, Donze M (1987) Anomalous behaviour of forward and perpendicular light scattering of a cyanobacterium owing to intracellular gas vacuoles. Cytometry 8:405–412

Dubelaar GBJ, Groenewegen AC, Stokdijk W, van den Engh GJ, Visser JWM (1989) Optical plankton analyser: a flow cytometer for plankton analysis. II. Specifications. Cytometry 10:529–539

Frankel DS, Olson RJ, Frankel SL, Chisholm SW (1989) Use of a neural net computer system for analysis of flow cytometric data of phytoplankton populations. Cytometry 10:540–550

Gerritsen J, Sanders RW, Bradley SW, Porter K (1987) Individual feeding variability of protozoan and crustacean zooplankton analyzed with flow cytometry. Limnol Oceanogr 32:691–699

Heiden T, Göhde W, Tribukait B (1990) Two-wavelength mercury arc lamp excitation for flow cytometric DNA-protein analyses. Anticancer Res 10:1555–1562

Hilton J, Rigg E, Jaworski G (1988) In vivo algal fluorescence, spectral change due to light intensity changes and the automatic characterisation of algae. Freshwater Biol 20:375–382

Hilton, J, Rigg E, Jaworski G (1989) Algal identification using in vivo fluorescence spectra. J Plankton Res 11:65–74

Hofstraat JW, van Zeijl WJM, Peeters JCH, Peperzak L, Dubelaar GBJ (1990) Flow cytometry and other optical methods for characterization and quantification of phytoplankton in seawater. SPIE 1269: Environment and Pollution Measurement Sensors and Systems, pp 116–132

Hutter KJ, Eipel HE (1979) Microbial determinations by flow cytometry. J Gen Microbiol 113:369–375

Hutter KJ, Eipel HE, Stohr M (1980) Flow cytometric analysis of microbial cell constituents after heavy metal intoxication. Acta Pathol Microbiol Scand (Suppl. 274):317–321

Lefort-Tran M, Bre MH, Pouphile M, Manigault P (1987) DNA flow cytometry of control *Euglena* and cell cycle blockade of vitamin B12 starved cells. Cytometry 8:46–54

Li WKW, Wood AM (1988) Vertical distribution of North Atlantic ultraphytoplankton: analysis by flow cytometry and epifluorescence microscopy. Deep Sea Res 35:1615–1638

Li WKW, Lewis MR, Lister A (1991) Picoplankton in the Gulf of Policastro. Signal and noise. Bigelow Laboratory Specialist Publication 4, part 2, p 3

Olson RJ, Frankel SL, Chisolm SW, Shapiro HM (1983) An inexpensive flow cytometer for the analysis of fluorescence signals in phytoplankton: chlorophyll and DNA distributions. J Exp Marine Biol Ecol 68:1–16

Olson RJ, Zettler ER, Anderson OK (1989) Discrimination of eukaryotic phytoplankton cell types from light scatter and autofluorescence properties measured by flow cytometry. Cytometry 10:636–643

Olson RJ, Chisholm SW, Zettler ER, Altabet MA, Dusenberry JA (1990) Spatial and temporal distributions of prochlorophyte picoplankton in the North Atlantic Ocean. Deep Sea Res 37:1033–1051

Ormerod MG (ed) (1990) Flow cytometry: a practical approach. IRL Press, Oxford

Paau AS, Oro J, Cowles JR (1978) Application of flow microfluorometry to the study of algal cells and isolated chloroplasts. J Exp Bot 29:1011–1020

Peeters JCH, Dubelaar GBJ, Ringelberg J, Visser JWM (1989) Optical plankton analyser: a flow cytometer for plankton analysis. I. Design considerations. Cytometry 10:522–528

Prezelin BB, Boczar BA (1986) Molecular bases of cell absorption and fluorescence in phytoplankton: potential applications to studies in optical oceanography. In: Round FE, Chapman DJ (eds) Progr Phycol Res 4:349–464

Price BJ, Kollman VH, Salzman GC (1978) Light scatter analysis of microalgae: correlation of scatter patterns from pure and mixed asynchronous cultures. Biophys J 22:29–36

Samson G, Popovic R (1988) Use of algal fluorescence for determination of phytotoxicity of heavy metals and pesticides as environmental pollutants. Ecotoxicol Environ Safety 16:272–278

Shapiro HM (1988) Practical flow cytometry, 2nd edn. Liss-Wiley, New York

Steen HB (1986) Simultaneous separate detection of low angle and large angle light scattering in an arc lamp-based flow cytometer. Cytometry 7:445–449

Trask BJ, van den Engh GJ, Elgershuizen JHBW (1982) Analysis of phytoplankton by flow cytometry. Cytometry 2:258–264

van Dilla MA, Dean PN, Laerum OD, Melamed MR (1985) Flow cytometry: Instrumentation and data analysis. Academic Press, New York

van de Hulst HC (1957) Light scattering by small particles. Wiley, New York

Vaulot D, Olson RJ, Chisholm SW (1986) Light and dark control of the cell cycle in two phytoplankton species. Exp Cell Res 167:38–52

Yentsch CM (1981) Flow cytometric analysis of cellular saxitoxin in the dinoflagellate *Gonyaulax tamarensis* var *excavata*. Toxicon 19:611–621

Yentsch CM (1990) Environmental health: flow cytometric methods to assess our water world. In: Darzynkiewicz Z, Crissman HA (eds) Flow cytometry. Academic Press, New York, pp 575–612

Yentsch CM, Horan PK (eds) (1989) Cytometry in aquatic sciences. Cytometry 10:497–669

Yentsch CS, Yentsch CM (1979) Fluorescence spectral signatures: the characterization of phytoplankton populations by the use of excitation and emission spectra. J Marine Res 37:471–483

Yentsch CM, Mague FM, Horan PK and Muirhead K (1983a) Flow cytometric determinations on individual cells of the dinoflagellate *Gonyaulax tamarensis* var *excavata*. J Exp Marine Biol Ecol 67:175–183

Yentsch CM, Horan PK, Muirhead K et al. (1983b) Flow cytometry and sorting: a technique for analysis and sorting of aquatic particles. Limnol Oceanogr 28:1275–1280

Chapter 11

Flow Cytometry at Sea

Glen A. Tarran and Peter H. Burkill

Introduction

The oceans constitute one of the world's largest resources, covering over 70% of the surface of the globe. They are known to play a major role in climate as the source of water for cloud formation and as a sink and transporter of heat absorbed from the sun's rays. It is now becoming clear that the oceans also have a biologically based role to play in climate through the cycling of carbon from the atmosphere to the ocean floor. Carbon dioxide (the major greenhouse gas) is drawn down from the atmosphere into the surface waters of the ocean by biological processes (Watson et al. 1991), where it is converted to organic carbon. Much of the organic carbon is recycled by the activity of the planktonic activity in surface waters. However, some of the organic carbon is lost from the surface and sinks to the sea bed where it ultimately becomes incorporated into the sediment. The biosynthetic conversion of carbon dioxide to organic carbon is carried out by the activity of single-celled plants, the phytoplankton. These cells are of microscopic size and are ubiquitous in the surface waters of the marine environment where they are often present in large numbers (Smetacek 1981; Porter et al. 1985; Fenchel 1988). Phytoplankton use carbon dioxide dissolved in the water to form proteins, carbohydrates and lipids through photosynthesis. This process uses chlorophyll to harvest sunlight, thereby providing the energy for photosynthetic reactions and so produces new organic material. Many phytoplankton are, in turn, eaten by the zooplankton, including other single-celled creatures, the protozoa. The importance of the single-celled phytoplankton and protozoa is enormous because although most are tiny (typically $<200\,\mu m$ in size), they account for almost 50% of the planet's biomass (Yentsch and Horan 1989). It has also been estimated that the phytoplankton transform between 2 and 4 gigatonnes of carbon dioxide to organic carbon every year (Takahashi 1989).

The importance of the plankton to climate and to marine food-web activity is only now beginning to be established, and research on this subject is highly topical. Our current knowledge of even fundamental questions about the community structure and variability of the plankton is poorly understood. Until recently, microscopy was the main tool available to scientists to analyse the plankton, their community structure and composition. The plankton was considered to be partitioned into discrete size classes (Sheldon et al. 1972). However, with the advent of particle counters such as Coulter counters it was realized that there was a continuum of particle sizes (Sheldon et al. 1972), the distribution of which varies greatly, both spatially and temporally. Coulter counters provided the first automated direct measurements of particle size distributions, and the rapidity with which they could analyse and size particles was an important development in the analysis of plankton populations.

Within the last decade marine science has made increasing use of flow cytometers. Although originally developed for biomedical research purposes, their capability for simultaneously measuring multiple parameters of single particles at high speed has proved invaluable to marine science in both laboratory studies (Campbell and Yentsch 1989; Cucci et al. 1985; Stoecker et al. 1986; Vaulot et al. 1986) and field studies (Campbell et al. 1989; Demers et al. 1989; Li 1989; Yentsch et al. 1986). The main advantage of flow cytometers over Coulter-type particle counters is that they are able to measure the fluorescence of particles. Since phytoplankton autofluoresce naturally, it is possible to differentiate these cell types readily from other particles in sea-water by flow cytometry. The analysis of phytoplankton is discussed in Chapter 10.

Cytometric analysis of oceanic planktonic populations can only be carried out through ship-board studies that involve taking the instrument to sea. The reason for this is that preservation of the fluorescence properties of phytoplankton is difficult, although some recent advances have been made (Vaulot et al. 1989). In addition, it is unknown how well some plankton groups are preserved, so at the current state of development it is an advantage not to manipulate the samples significantly before analysis.

Since 1986, the Plymouth Marine Laboratory (PML) of the United Kingdom's Natural Environment Research Council (NERC) has undertaken flow cytometry work at sea on several cruises. These have involved using a Coulter EPICS 741 flow cytometer to investigate phytoplankton dynamics. Many logistical and technical problems had to be solved before being able to work successfully at sea. This chapter describes the logistics involved in taking and using a large flow cytometer at sea and illustrates the nature and areas in which the work has been undertaken.

The Flow Cytometry Container Laboratory

The Coulter EPICS 741 flow cytometer is large (3 m long × 1 m wide × 1.5 m high), heavy (300 kg) and complex. It requires three-phase and single-phase electricity, a pressurized water cooling system for the laser and a bottled nitrogen supply. These requirements make housing the EPICS within the internal laboratory space of research ships very difficult. To carry

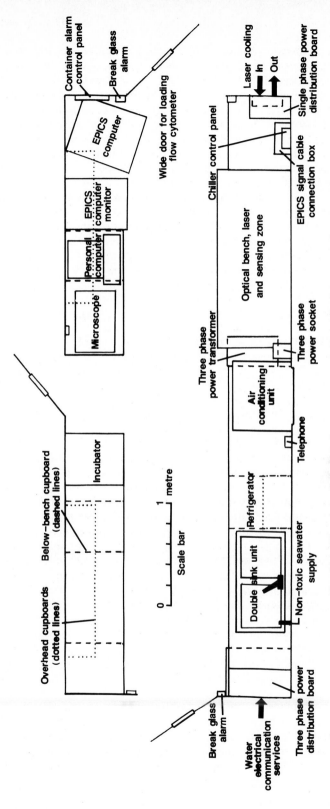

Container alarm control panel
Break glass alarm
EPICS computer
EPICS computer monitor
Personal computer
Microscope
Wide door for loading flow cytometer
Laser cooling
In
Out
Single phase power distribution board
Chiller control panel
EPICS signal cable connection box
Optical bench, laser and sensing zone
Three phase power socket
Three phase power transformer
Air conditioning unit
Telephone
Refrigerator
Double sink unit
Non-toxic seawater supply
Break glass alarm
Water electrical communication services
Three phase power distribution board
Overhead cupboards (dotted lines)
Below-bench cupboard (dashed lines)
Incubator
Scale bar
0
1 metre

Fig. 11.1. Plan view of the PML flow cytometry container laboratory set up for research at sea.

out ship-board research a customized container has been used to provide the necessary laboratory space and interfacing to the services required. After the first two cruises in which the EPICS was taken to sea in borrowed containers, a purpose-built container specifically designed for flow cytometry was funded by NERC's Biogeochemical Ocean Flux Study (BOFS). The container (6 m long × 2.4 m wide × 2.6 m high) was designed by NERC's Research Vessel Services and built by a local contractor. After testing to Lloyd's regulations, the cytometer was fitted with tracking rails set into the walls and floor to fix equipment, lighting, and the necessary service connections for NERC research vessels (laser cooling system, sea-water, hot and cold freshwater, three-phase and single-phase power connections plus ship's 'phone and alarm). The container was also fitted with an air-conditioning unit to dissipate the 4 kW of heat generated by the EPICS and other equipment. The container was then delivered to the Plymouth Marine Laboratory where it was customized further with benching, incubators, sinks and refrigerator (Fig. 11.1) to act as a mobile laboratory for ship-board flow cytometry.

Mobilization for a Cruise

Successful ship-board research depends upon meticulous preparation before the cruise. Before going to sea there are many operations that must be carried out. The EPICS must be moved from the laboratory within PML to the cytometer container situated outside. To help with this problem and to reduce ship-board vibration, shock-absorbing trolleys for the EPICS laser and computer units were designed and built in the laboratory's workshops (the EPICS remains on them permanently, even in the laboratory). To stabilize the EPICS laser unit, a special heavy-duty wooden bench, over 5 cm thick, was constructed and this was attached permanently to the EPICS pedestal units. The EPICS requires five people to move it into the container because of its size and weight. The trolley units are designed to bolt directly to the tracking in the container's floor. Temporary three-phase and single-phase electricity cables are then connected and the laser cooling chiller coupled up to test the EPICS prior to transportation. All other necessary equipment (personal computer, microscope, centrifuge and consumables) are installed and securely fastened to stop them moving while at sea.

When loaded, the container and the chiller weigh c. 7 tonnes. These require an industrial crane to hoist them onto a lorry for transportation to the port where the research vessel is docked. At the ship, the container and chiller are then lifted into position on board (Fig. 11.2). The container is bolted down and the chiller is welded to a steel flat-bed to secure it. The main's services from the ship are connected to the container by the ship's or dockside engineers. The laser cooling system is tested and the laser electronically tuned to the ship's power supply by a Coulter engineer who also services the EPICS prior to departure. A selection of engineering spares (solenoids, relays, etc.) provided by Coulter are also carried for

Fig. 11.2a,b. The flow cytometry container: **a** being loaded onto RRS *Charles Darwin*; **b** showing the EPICS cytometer inside the container.

emergency use at sea. Following these operations, the EPICS is ready for ship-board use.

After the cruise, there is a reversal in the above activities. After reconnection to the laboratory services, and a further visit by the Coulter engineer, the system is available for normal use in the laboratory.

Table 11.1. The precision of ship-board cytometric analysis can be determined by comparison of results of laboratory and ship-board tests of light scatter and fluorescence derived from fluorosphere reference beads

Date	FALS[a]	IGFL[b]	Comments
Laboratory-based results: 1986			
21 March	1.53	2.24	
8 April	2.20	2.17	
18 April	1.58	2.3	
Average	1.77	2.23	
Ship-based results: RRS Charles Darwin 1986			
30 May	1.82	2.28	8 knot wind; hove to
31 May	1.71	2.41	16 knot wind; hove to
1 June	1.95	2.73	11 knot wind; hove to
2 June	1.96	2.56	9 knot wind; hove to
3 June	2.96	3.56	17 knot wind; steaming at 13 knots
4 June	2.1	2.75	16 knot wind; hove to
Average	1.90	2.54	Excluding 3 June

[a] FALS, narrow angle forward light scatter.
[b] IGFL, integral green fluorescence.

Operations and Scientific Studies at Sea

While at sea, daily checks are made on the stability and alignment of the laser and optics using standard calibration beads. The EPICS performs remarkably well under a variety of weather conditions, as shown in Table 11.1. The demonstration of the reliability of the alignment is very important, especially at sea where the EPICS is subjected to movement for which it was not designed.

As ship time is limited and oceanographic samples are valuable, as much data as possible is obtained from each sample. This situation differs from most laboratory situations where experimental conditions are controlled and analyses could, if necessary, be repeated. As a result, most of the data are stored on "list-mode" on the 1 Mb 8 inch disk and analysed subsequently back in the laboratory. On a typical 3–4 week cruise, the EPICS may be run for as long as 400 h, amassing a data set consisting of 200 × 1 Mb diskettes with up to four list-mode data files on each disk.

The EPICS flow cytometer at PML first underwent sea-trials in 1986 and has since participated in five cruises (Table 11.2). All cruises have been in the North Atlantic and North Sea in the regions shown in Fig. 11.3 and have involved the EPICS instrument on all but one cruise. In 1989, a Coulter PROFILE was used on board the German ship FV *Meteor*. While the initial cruises were primarily concerned with assessing ship-board capabilities and design optimization, the main scientific focus has been with the Biogeo-chemical Ocean Flux Study (BOFS). The goal of BOFS has been to quantify reservoirs and fluxes of carbon in the ocean and the role played by plankton in oceanic cycling of carbon.

Table 11.2. Ship-board cytometric operations carried out with the PML EPICS 741 cytometer between 1986 and 1991

Ship	Region (see Fig. 11.3 for cruise tracks)	Time period	Objectives
RRS *Charles Darwin*	Celtic Sea (1)	June 1986	Sea-trial of EPICS 741
RRS *Challenger*	North Sea (2)	August 1987	Picoplankton sorting
FV *Meteor*	Tropical Atlantic (3)	March 1989	BOFS: latitudinal studies
RRS *Discovery*	NE Atlantic (4)	July 1989	BOFS: latitudinal studies
RRS *Charles Darwin*	NE Atlantic (5)	June 1990	BOFS: lagrangian study
RRS *Charles Darwin*	Iceland Basin (6)	June 1991	BOFS: coccolithophore biogeochemistry

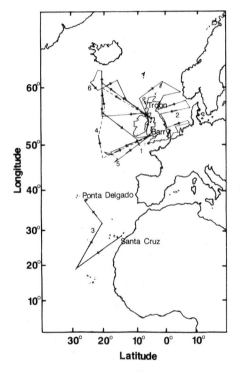

Fig. 11.3. Chart showing the cruise tracks upon which ship-board flow cytometry research has been carried out between 1986 and 1991.

The main objective of BOFS ship-board flow cytometry has been to analyse the distribution and dynamics of phytoplankton in the ocean. The analysis of phytoplankton depends upon the autofluorescence signatures given by chlorophyll when irradiated by blue light. To analyse the phytoplankton cytometrically, the 488 nm line of the argon ion laser is used to irradiate water samples. These are obtained from discrete depths in the ocean by water bottles that are triggered remotely at depth. Samples are

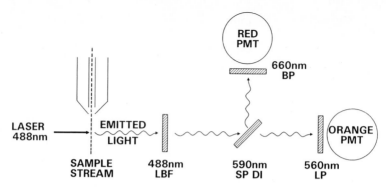

Fig. 11.4. Optical protocol of the analysis of phytoplankton using the EPICS 741 cytometer. PMT, photomultiplier tube; LBF, laser blocking filter; SP, short-pass filter; LP, long-pass filter; BP, band-pass filter; DI, dichroic mirror.

pre-screened through $50\,\mu$m mesh to remove large particles that might clog the instrument and are then analysed cytometrically. The standard optical configuration for the analysis of phytoplankton is shown in Fig. 11.4. The cellular parameters analysed are log forward angle light scatter (LFALS) for particle sizing, log integral red fluorescence (LIRFL) derived from chlorophyll and log integral orange fluorescence (LIOFL) derived from phycoerythrin.

Data on particle size are collected because they can be related to cellular biomass. Information on chlorophyll and phycoerythrin fluorescence is collected because it allows characterization of the phytoplankton and differentiation from non-phytoplankton particles in sea-water. As chlorophyll is present in all algae its autofluorescence is used as a general cytometric criterion for classification of phytoplankton. However, phycoerythrin is restricted to cryptomonads and almost all oceanic cyanobacteria and so forms a marker for these phytoplankton taxa. Logarithmic signals are used in the analysis because the size and pigment content can vary by more than two orders of magnitude. The size range of interest is typically between 2 and $20\,\mu$m and spans a size range known as the nanoplankton (Sieburth et al. 1978).

Analysis of data is normally carried out on two-parameter scatter plots. Fig. 11.5 illustrates a vertical profile of samples obtained from between 10 and 100 m at 33° N 220° W in the North Atlantic in April 1989. Each plot shows log chlorophyll plotted against log phycoerythrin to analyse and differentiate phytoplankton with and without phycoerythrin. The different components of the population are then gated (represented as boxes in Fig. 11.5) and their numbers calculated using the cytometer's computer software. An analysis of these data is shown in Fig. 11.6.

Fig. 11.6 shows the abundance of total phytoplankton together with those phytoplankton with and without phycoerythrin. Between 10 m and 20 m the numbers of phytoplankton are relatively low and they have low cellular fluorescence (Fig. 11.5). However, between 40 m and 70 m the numbers of phytoplankton containing just chlorophyll (near the abscissa on Fig. 11.5) and those with proportionally similar chlorophyll and phycoerythrin contents

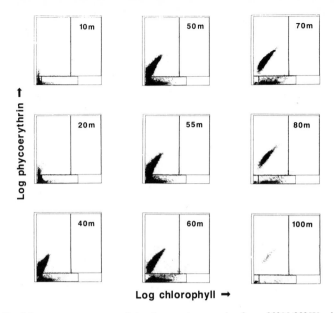

Fig. 11.5. Dual fluorescence scatter plots of seawater samples from 33° N 22° W, showing how chlorophyll-containing and phycoerythrin-containing organisms vary with depth. Samples were taken aboard FV *Meteor* in April 1989.

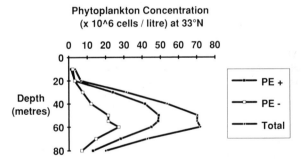

Fig. 11.6. Vertical distribution of total phytoplankton and phytoplankton with and without phycoerythrin from 33° N 22° W. PE+, phytoplankton containing both phycoerythrin and chlorophyll; PE−, phytoplankton containing only chlorophyll.

(at approximately 45° on Fig. 11.5) increase significantly. A sub-surface chlorophyll maximum is evident at 60 m, whereas the phycoerythrin maximum was situated at 50 m. The vertical displacement of the two chromatic groups of phytoplankton is probably due to preference for different light conditions, which, at such depths will vary significantly over a short distance.

It is known that the spring phytoplankton bloom in the temperate North Atlantic initiates at low latitudes and progresses to higher temperate latitudes during the spring and summer. Fig. 11.7 illustrates the size and fluorescence distributions of phytoplankton taken 4 days apart in the Iceland Basin at

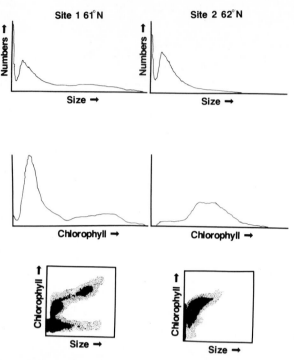

Fig. 11.7. Distributions of phytoplankton and other particles along the 20° W meridian at 61° N and 62°19′ N in July 1991. *Upper*, size distributions; *middle*, chlorophyll distributions; *lower*, scatter plots of size versus chlorophyll.

61° N and 62° N during July 1991. They represent a post-bloom area (61° N) and an actively blooming area (62° N). The particle size distributions at 61° N and 62° N are very similar (Fig. 11.7, upper), no major differences in the communities being apparent. However, when the chlorophyll distributions are analysed (Fig. 11.7, middle), two different communities are revealed. The two parameter size and fluorescence scatter plots obtained by cytometric analysis provide further information about the communities (Fig. 11.7, lower). At 61° N the size range of phytoplankton is much greater than at 62° N. There are also large numbers of particles with low levels of fluorescence at 61° N that are absent at 62° N. The wide range of phytoplankton sizes and the presence of many low-fluorescence particles is characteristic of communities after a phytoplankton bloom and through into the summer. A smaller size range and absence of low-fluorescence particles is more typical of an area in which phytoplankton is actively blooming.

During 1989 investigations were made within the BOFS programme to quantify the patterns of vertical distribution of the phytoplankton at different latitudes in the North Atlantic. The results from a selection of stations are shown in Fig. 11.8. The samples at 60° N and 47° N were taken during July after the spring bloom, whereas the samples at 33° N and 18° N were taken during the spring bloom in March. Phytoplankton abundance was higher at the low-latitude spring bloom stations than at northerly latitudes

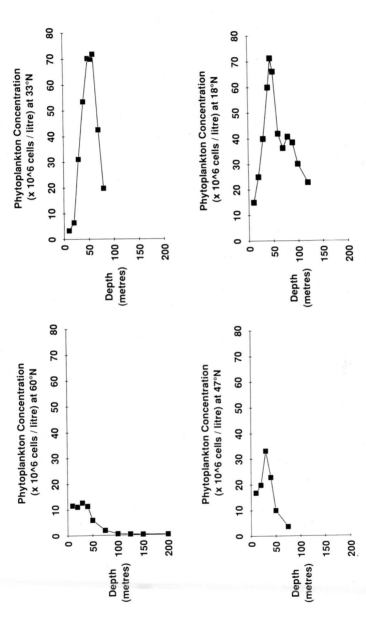

Fig. 11.8. Vertical profiles of phytoplankton abundance at four latitudes in the North Atlantic in 1989. Data obtained as part of the UK Biogeochemical Ocean Flux Study.

experiencing summer, post-bloom conditions. At all stations phytoplankton concentrations decreased markedly with depth, particularly below the sub-surface chlorophyll maximum situated at 40–60 m. It can be seen that it is possible to gain much information about the abundance and distribution of nanoplankton from cytometric analysis. This information was not so readily obtained before flow cytometers were taken to sea, as alternative methods of analysis rely either on microscopic analysis of preserved samples or on a combination of Coulter counting and high performance liquid chroma-tography to quantify particulates and pigments respectively. Both methods involve analysis of samples which take, perhaps, one or two orders of magnitude more time than cytometric analysis to obtain the data required.

The dynamics of phytoplankton populations have also been investigated cytometrically at sea. The phytoplankton form a significant food resource and investigations have also been carried out on their mortality from protozoa and other microzooplankton grazers. The microzooplankton are phagotrophic organisms of microscopic size that are typically present at concentrations >1000 per litre. To quantify the mortality rates of phyto-plankton due to grazing and their growth rates, dilution experiments based on the concept of Landry and Hassett (1982) are carried out. This assumes that while phytoplankton growth rates are density independent, their rates of mortality are density dependent. Time course incubations of the natural microplankton assemblage at different dilutions are made up with filtered particle-free sea-water to experimentally decouple the density-dependent grazing mortality of phytoplankton from their growth. Phytoplankton con-centrations are determined cytometrically and the cell concentrations inserted into a simple mathematical model to determine the specific growth rate of the phytoplankton at the different dilutions. The results of a dilution experiment to quantify the grazing mortality and growth rates of phyto-plankton are shown in Fig. 11.9.

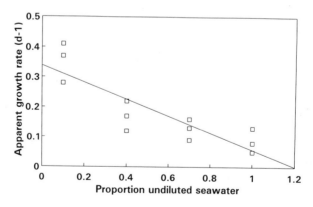

Fig. 11.9. Phytoplankton dynamics determined cytometrically using the dilution technique. The water sample was obtained from 10 m depth at 60° N 20° W on 19 June 1991. In this experiment three replicate samples were incubated and the phytoplankton quantified cytometrically, initially and after 24 h. The phytoplankton mortality rate due to microzooplankton grazing (indicated by the gradient of the data) was 0.28 per day while the phytoplankton growth rate (indicated by projected intercept of data on the ordinate) was 0.34 per day.

In this experiment, the specific mortality rate of phytoplankton is 0.28 per day while the specific growth rate is 0.34 per day. Three points are note-worthy: first, the microzooplankton are grazing much of the phytoplankton each day; second, growth and mortality are quantitatively similar (and so the net change in the phytoplankton population is low); and third, the rates of both growth and grazing are high. These rates typify the plankton com-munity during the summer months and show that the plankton community is dynamically active.

Future Developments

In the past few years flow cytometry has made significant contributions to marine science (Burkill 1987) and much of this has been through ship-board operations. Yet the cytometers that are used in ship-board work are less than ideal for many marine applications. Large cytometers are bulky and difficult to handle; future usage of flow cytometers at sea should ideally use smaller machines. Some of these are already commercially available while others have been built specifically for phytoplankton research (Cunningham 1990; Dubelaar et al. 1989; Peeters et al. 1989). The advantages of using small machines are obvious. They require less room and could be housed in the normal laboratory space in the ship. Their smaller size also eases and cheapens transportation. The recent improvements in cytometer optics allow the use of small air-cooled lasers or arc lamps that avoid the requirement of complex cooling systems. This in turn avoids the requirements for three-phase power, so allowing use of the new generation of cytometers on smaller simpler ships. In spite of this, it is possible that commerical cytometers may not be suitable for some specific marine applications. In this respect, development of specialized flow cytometers may be necessary (Dubelaar et al. 1989; Peeters et al. 1989). The final consideration is that of cost: mobilization of a large flow cytometer can cost thousands of pounds and is therefore a very expensive process regarding both money and man-power.

Biomedical research flow cytometers are designed to operate with samples containing high particle concentrations, typically $10^5 - 10^6$ particles/ml. With the particle sizes used in our ship-board studies particle concentrations are normally between $10^3 - 10^4$ particles/ml. Sample flow rates are usually between 20 and $50 \mu l$/min and quantitative volume measuring devices attached to commercial flow cytometers measure up to $250 \mu l$. These flow rates and sample sizes have meant that to analyse a sample and get a statistically relevant count has, in the past, taken up to 30 min per sample. Recently, Casotti et al. (1990) and ourselves (unpublished results) have increased the rate of sample analysis by nearly two orders of magnitude through redesigning the sample delivery system of the EPICS cytometer. Such modifications have been invaluable in phytoplankton research, espe-cially for quantifying groups such as coccolithophorid phytoplankton that may be found in low (100–400 cells/ml) but biogeochemically significant concentrations. Further refinement of this technique is likely to be of use particularly for resolving larger, less abundant organisms, as well as of general use for the study of natural plankton populations.

As the interest in using flow cytometry in marine science increases, the size range of organisms that is studied increases. Although sub-micrometre sized particles such as marine bacteria (Robertson and Button 1989), prochlorophytes (Chisholm et al. 1988) and cyanobacteria (Olson et al. 1988) can be resolved, and particles as large as fish larvae several millimetres in length can be sized and counted (Hüller et al. 1991) using very different machines, there is, as yet no flow cytometer available for analysing the wide size range of particles present in sea-water. At the current state of marine cytometric development, a narrow window of particle size (0.5–5, or 2–20 μm, etc.) can be analysed on any one occasion. Increasing this is only possible by time-consuming reanalysis of the sample with different cytometer settings.

Flow cytometry as a taxonomic tool in plankton research is a new and exciting area for future development. Although there have been successes in identifying certain groups of plankton in the laboratory (Olson et al. 1989) and in the field (Robertson and Button 1989; Casotti et al. 1990; Chisholm et al. 1988) the challenge remains to differentiate even the well-known groups such as dinoflagellates and diatoms in mixed natural assemblages. However, much work is now under way to develop species- and group-specific fluorescent oligonucleotide probes. A probe is designed to be incorporated into unique DNA sequences of the targeted organism. As the probe is fluorescent the target organism can be differentiated from other organisms by its fluorescence signature. These molecular techniques have the potential not only for separating different groups and species of phyto-plankton, but also for the characterization of heterotrophic and mixotrophic plankton organisms. Advances of this kind could be of enormous value in automating plankton analysis at sea in the future.

In conclusion, flow cytometry is providing marine science with a very powerful and useful tool for carrying out marine research in the field that could not be done before. There is huge potential for developing flow cytometry into a standard ship-board analytical technique for making rapid quantitative determinations of the different components of plankton communities.

Acknowledgements

We thank the NERC Research Vessel Services in Barry for designing and building the cytometry container, David Robins and workshop staff at PML for fitting out the container, and Coulter Electronics (Luton) for their support of our research. The research forms an integral part of both the UK Biogeochemical Ocean Flux Study and the PML Laboratory Research Project 2.

References

Burkill PH (1987) Analytical flow cytometry and its application to marine microbial ecology In: Sleigh MA (ed) Microbes in the sea. Ellis Horwood, Chichester

Campbell JW, Yentsch CM (1989) Variance within homogeneous phytoplankton populations. II. Analysis of clonal cultures. Cytometry 10:596–604

Campbell JW, Yentsch CM, Cucci TL (1989) Variance within homogeneous phytoplankton populations. III. Analysis of natural populations. Cytometry 10:605–611

Casotti R, Olson RJ, Zettler ER (1990) Sea-going flow cytometric discrimination of coccolithophorids. In: Individual cell and particle analysis in oceanography. Abstracts of a NATO Advanced Study Institute, Aquafredda di Maratea, Italy, 21–30 October, 1990

Chisholm SW, Olson RJ, Zettler ER, Goericke R, Waterbury JB, Welschmeyer NA (1988) A novel free-living prochlorophyte, abundant in the oceanic euphotic zone. Nature 334:340–343

Cucci TL, Shumway S, Newell RD, Selvin R, Guillard RRL, Yentsch CM (1985) Flow cytometry: a new method for the characterisation of differential ingestion, digestion and egestion by suspension feeders. Marine Ecol Prog Ser 24:201–204

Cunningham A (1990) Flow diffractometry of phytoplankton cells: a simple optical approach to rapid size and shape discrimination. In: Individual cell and particle analysis in oceanography. Abstracts of a NATO Advanced Study Institute, Aquafredda di Maratea, Italy, 21–30 October, 1990

Demers S, Davis K, Cucci TL (1989) A flow cytometric approach to assessing the environmental and physiological status of phytoplankton. Cytometry 10:644–652

Dubelaar GBJ, Groenewegen AC, Stockdijk W, van den Engh GJ, Visser JWM (1989) Optical plankton analyser: a flow cytometer for plankton analysis. II. Specifications. Cytometry 10:529–539

Fenchel T (1988) Marine plankton food chains. Annu Rev Ecol Syst 19:19–38

Hüller R, Püffgen W, Gloßner E, Hummel P, Kachel V (1991) A macro flow planktometer for the analysis of large plankton organisms. Cytometry (Suppl.) 5:53

Làndry MR, Hàssett RP (1982) Estimating the grazing impact of marine microzooplankton Marine Biol 67:283–288

Li WKW (1989) Ship-board analytical flow cytometry of oceanic ultraphytoplankton. Cytometry 10:564–579

Olson RJ, Chisholm SW, Zettler ER, Armburst EV (1989) Analysis of *Synechococcus* pigment types in the sea using single and dual-beam flow cytometry. Deep Sea Res 35:425–440

Peeters JCH, Dubelaar GBJ, Ringelberg J, Visser JWM (1989) Optical plankton analyser: a flow cytometer for plankton analysis. I. Design considerations. Cytometry 10:522–528

Porter KG, Sherr EB, Sherr BF, Pace M, Sanders RW (1985) Protozoa in planktonic food-webs. J Protozool 32:409–415

Robertson BR, Button DK (1989) Characterising aquatic bacteria according to population, cell size and apparent DNA content by flow cytometry. Cytometry 10:70–76

Sheldon RW, Prakash A, Sutcliffe WH Jr (1972) The size distribution of particles in the ocean. Limnol Oceanogr 17:327–340

Sieburth J McN, Smetacek V, Lenz J (1978) Pelagic ecosystem structure: heterotrophic compartments of the plankton and their relation to plankton size fractions. Limnol Oceanogr 23:1256–1263

Smetacek V (1981) The annual cycle of protozooplankton in the Kiel Bight. Marine Biol 63:1–11

Stoecker DK, Cucci TL, Hulbert EM, Yentsch CM (1986) Selective feeding by *Balanion* sp. (Ciliata: Balanionidae) on phytoplankton that best support its growth. J Exp Mar Biol Ecol 95:113–130

Takahashi T (1989) The carbon dioxide puzzle. Oceanus 32:22–29

Vaulot D, Olson RJ, Chisholm SW (1986) Light and dark control of the cell cycle in two marine phytoplankton species. Exp Cell Res 167:38–52

Vaulot D, Courties C, Partensky F (1989) A simple method to preserve oceanic phytoplankton for flow cytometric analysis. Cytometry 10:629–635

Watson AJ, Robinson C, Robertson JE, Williams PJ LeB, Fasham MJ (1991) Spatial variability in the drawdown of atmospheric carbon dioxide in the North Atlantic, Spring 1989. Nature 350:50–53

Yentsch CM, Horan PK (1989) Cytometry in the aquatic sciences. Cytometry 10:497–499

Yentsch CM, Cucci TL, Phinney DA, Topinka J (1986) Real-time characterisation of individual marine particles at sea: flow cytometry. In: Bowman M, Yentsch CM, Peterson WT (eds) Tidal mixing and plankton dynamics. Lecture notes on coastal and estuarine studies no. 17. Springer, New York, pp 414–448

Chapter 12

Neural Network Analysis of Flow Cytometry Data

Lynne Boddy and Colin W. Morris

Introduction

Flow cytometry can measure several variables (typically between three and eight) on cells very rapidly (often in excess of 10^3 per second). Hence it is a major challenge to analyse the data produced. Currently, we are aware of four published studies employing neural computing techniques to distinguish between different cell types on the basis of flow cytometry data: Frankel et al. (1989) and Balfoort et al. (1992) analysed phytoplankton populations, Noehte et al. (1989) classified normal and aberrant chromosomes, and Morris et al. (1992) analysed fungal spore populations. This chapter concerns the potential use of neural computing networks to analyse flow cytometry data.

What is Neural Computing and What Are Neural Networks?

Neural computing seeks to model some of the attributes of the brain – particularly the powerful thinking, storage and problem-solving capabilities (e.g. McClelland et al. 1986; Rumelhart et al. 1986; Boddy et al. 1990; Dayhoff 1990). The fundamental cellular element of the brain is the neuron, and there are about 10^{11} of these in the human brain. In simple terms, a neuron consists of dendrites which receive inputs from other neurons, the main body of the cell and the axon – a longitudinally extended projection which divides to form several endings which connect with dendrites of other neurons (Fig. 12.1). Output signals, in the form of short pulses of electrical activity at about 1000 Hz, pass along the axon to be transmitted to other cells. Neurons are not directly connected with each other but have a gap – the synapse – between dendrites and axons. Signals traverse the gap as chemicals (neurotransmitters). The strength of the signal at the dendrites

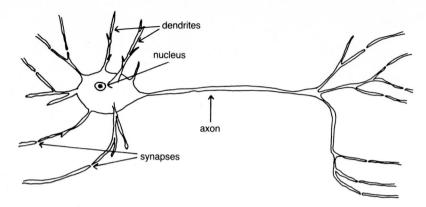

Fig. 12.1. Representation of a human neuron.

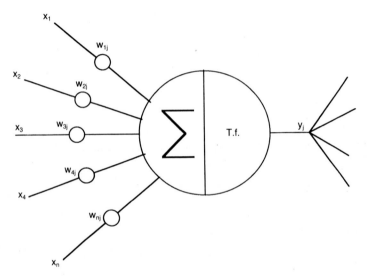

Fig. 12.2. A node or processing element, with inputs (x_i) being weighted (w_{ij}), and an output y_j. Summation of the weighted elements occurs within the node-$\Sigma w_{ij}x_i$. A transfer function (T.f.) is applied to the summation. See text for details.

depends on the amount of chemical released (which is directly proportional to the frequency of nerve impulses arriving at the synapse) and the membrane potential of the dendrite (which is directly proportional to the amount of chemical received). Excitatory and inhibitory events at the synapse, plus a combination of information from all dendrites in neurons, provide the basic memory mechanism.

The building blocks of neural computing networks (termed nodes or processing elements) are analogous to neurons (Fig. 12.2). Many inputs enter the node where they are then mathematically combined, usually by a simple summation of the weighted inputs. The combined value is then trans-

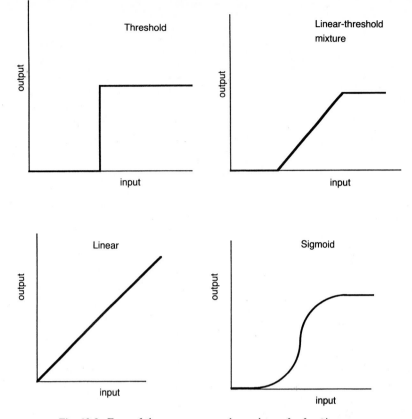

Fig. 12.3. Four of the more commonly used transfer functions.

formed by a transfer function which passes on information to subsequent nodes, and is equivalent to firing a neuron. The most commonly employed transfer functions are sigmoidal, having a general form $1/[1 + e^{-s}]$, where s is the sum of the weighted inputs to a node (Fig. 12.3). Other transfer functions, e.g. threshold, linear, and combined threshold and linear, can be adopted (Fig. 12.3).

As in the brain, individual nodes are interconnected to form a neural network (Fig. 12.4). Most neural networks consist of a number of layers of nodes: an input layer, an output layer and one or more hidden layers between these. Connections between nodes vary in different architectures. In some types of neural computing networks the nodes in one layer connect only with the nodes in the next layer, while in other types of network there can also be interconnection between nodes in the same layer and connection back to nodes in previous layers. Connections can be complete, partial or random.

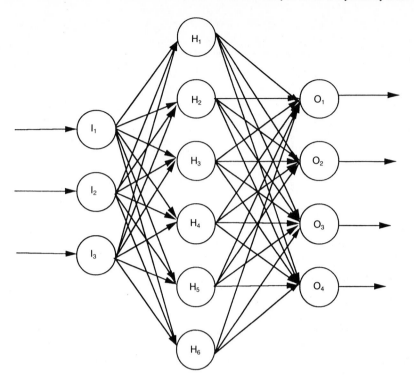

Fig. 12.4. Representation of a simple neural network having three input nodes (I_1–I_3), six nodes in one hidden layer (H_1–H_6) and four output nodes (O_1–O_4). All inputs and outputs from hidden nodes are weighted (see text for details).

Training Neural Computing Networks

Unlike traditional computer systems, neural networks do not follow explicit rules in the form of a program, rather they "learn" by being shown examples. There are a variety of different ways by which neural computing networks can be trained, which fall under two broad headings: supervised and unsupervised training/learning. With unsupervised learning the network is presented with inputs and allowed to organize itself so as to formulate its own classifications (see Rumelhart et al. 1986). Such methods may be useful in taxonomy for picking out groups with similar characteristics. For identification purposes, however, supervised learning is probably more appropriate. This involves presenting the neural network with input and known output data, and adjusting the weightings to give the required output for a given input. There is a range of methods for doing this (Dayhoff 1990; Forsyth 1990; NeuralWare Inc. 1991), but considerable work remains to be done in order to determine which is the most appropriate method to use under different circumstances. Currently, the most commonly used supervised training method is back-error propagation (often called simply back-propagation) (e.g. Dayhoff 1990), which employs an architecture where nodes in one layer connect only with nodes in the following layer (i.e. feedforward). A training set of data is constructed, consisting of a number of training patterns of known inputs (measurements of flow cytometry parameters for example) and outputs (type or species upon which flow cytometry measurements were made). Initially a single pattern of inputs is applied to a network having random weightings, and output is produced. The latter will differ from the required output, and an error value is computed by comparison of actual with required output. Then, weightings for the output elements are slightly modified so as to reduce the error value. Error values and alterations to weightings are determined by the Delta (or Widrow-Hoff) rule, which can be expressed as:

$$\Delta w_{ij} = \varepsilon e_j y_j$$

and

$$e_j = r_j - y_j$$

where Δw_{ij} is the required weight change for input i to node j, e_j is the error in the output for that input, ε is a learning rate constant, and y_j and r_j are the actual and required outputs respectively. This process is repeated for each layer going backwards through the network. The whole operation is then repeated for the next pattern in the training set, employing the modified weightings produced by the previous presentation of a pattern. This is continued, working through all of the training patterns in turn. Alternatively, each pattern from the training set can be presented to the network with random weights and the new weights determined; a cumulative value of the weightings at each input to a node is then obtained, and updating of weightings occurs only after all patterns in the training set have been presented. For successful learning, i.e. to achieve observed outputs close to known/actual outputs, numerous presentations of the training set are usually required. A measure of how well a network has learned the data is provided by the sum of squares of the difference (SSD) between the actual and expected output for the training data. Training is usually allowed to continue until a suitably low SSD has been reached. To determine how well the trained network can generalize or predict from other data, an "unseen" test set of input patterns, with known output patterns, can be presented to the trained network and the percentage of correctly predicted output patterns determined.

Training can be based on a sample of "list mode" flow cytometer data such that the data on any one particle form one training pattern. Alternatively, the *distributions* of measured variables for an entire sample could form a single training pattern, necessitating data from numerous samples to provide a training set. The latter approach is suitable if predictions of identity are always to be made from pure samples, but not if mixtures of different types of particle are likely to be used. The former approach is suitable for mixtures of different types of particles and for pure cultures. Moreover, it is probably more amenable for on-line real-time analysis (see below).

Neural network software usually requires that input values are normalized so that they range between 0 and 1; for "list mode" data this simply involves dividing the raw data for each parameter by the number of channels of the flow cytometer (commonly 255 or a multiple of it). Also, if measurements are made with different gain settings of the flow cytometer, further division of input values is required to maintain the same scale. Alternatively, it may prove useful to use gain setting as an network input parameter (see below).

Values at output nodes are continuous variables between 0 and 1 with most software. When outputs are of qualitative/attribute data, such as those from flow cytometry where particle type (e.g. species) is the output parameter, the values of output nodes must form a numerical code, with a different code for each species. For example, if a network was being trained to distinguish between four different species, a value of 1 at the first output node and 0 at another three (i.e. 1,0,0,0) could code for species A; 0,1,0,0 could code for species B; and 0,0,1,0 and 0,0,0,1 could code for species C and D respectively. In practice, however, since output values are continuous variables they are rarely 0 or 1 but something between, and it is convenient to take the largest value as being equivalent to 1 in the code and the other values as being equivalent to 0. Thus, for example, an output pattern of 0.6921, 0.3121, 0.2987, 0.2248 would be treated as being 1,0,0,0 which codes for species A. Output values can be loosely thought of as reflecting the "closeness-of-fit" of the identification; thus low values inspire less confidence than high values.

Neural Network Software and Hardware Platforms

Early neural network research required that all workers wrote their own software for experimentation, but during the last 5 years a number of neural network packages have become commercially available, and some of these are mentioned below. This area is expanding all the time, hence what is written in this section is likely rapidly to become outdated.

McClelland and Rummelhart (1988) provide software, covering a number of neural networks applications, with their book. The back-propagation implementation is suited to small problems and allows the user to experiment by changing virtually every parameter possible in the model. The major drawback with the software is the idiosyncratic user interface, which requires several files to be set up to define the network, the screen format and its training patterns. The format of these files may be rather daunting to the novice user.

NeuralWare Inc. produce high-quality packages (NeuralWorks Explorer and NeuralWorks Professional II/PLUS – the latter being the more comprehensive and allowing larger networks to be constructed) which are easy to use and allow the user to interact with the package to design networks and make changes as required. The packages are very flexible and support a number of different network paradigms. They are suited to those problems requiring large networks or where the user wishes to investigate the use of different paradigms. The NeuralWorks Explorer package allows back-propagation networks to be configured with a maximum, total number of nodes (i.e. input, hidden and output nodes) of 150.

Other packages include Brainmaker (California Scientific Software, distributed in the UK by Tubb Research Ltd., Hampshire), which is a user-friendly back-propagation implementation; Braincell (Promised Land

Technologies Inc., New Haven, CT, USA), designed to run under the Microsoft Excel spreadsheet package and Windows; and NeuroWindows (Ward Systems Group Inc., Frederick, MD, USA), which is designed to integrate with Microsoft Visual Basic to produce Windows neural networks implementations.

The above packages all run on IBM PC and compatible machines, but some have also been ported to other platforms; for example, the NeuralWare Inc. software is available for Apple Macintosh and SUN SPARC stations. When running neural network simulation packages the computer needs to be as fast as possible so that the processing of the essentially parallel architecture of the neural network can be achieved as quickly as possible. It is recommended that maths co-processors are used wherever possible as this will give a great increase in processing speed.

Large neural net applications will require a large amount of processing time on a conventional computer system and special hardware may be required to make the application run in a reasonable timescale. The options available are highly parallel machines such as Transputer arrays or the use of dedicated neural network chips which run the neural network in hardware. The latter approach is very promising and a review of currently available neural network chips is provided by Schwartz (1990).

Training Neural Computing Networks on Flow Cytometry Data: Parameters that May Affect Learning

There are a number of parameters within the neural network software that affect learning. In the first instance it is as well to leave these at their default settings, as these are usually good first guesses. However, some network parameters (such as number of hidden nodes) and external parameters (such as training cycles or epochs, i.e. complete presentations of the training set, and number of different training patterns) affect speed of learning, how well a network learns and, indeed, the ability to learn at all. These parameters have to be experimented with to optimize neural network configuration and learning.

Networks with the lowest possible number of nodes giving successful learning should usually be selected, to ensure that the network "can make generalizations" rather than becoming specific to the training patterns. Various attempts have been made to predict the number of hidden nodes required for any particular implementation, one such rule of thumb estimating the required number as: $2\times$ (number of input nodes + number of output nodes)$^{0.5}$ (Anon 1991).

In the study on identification of fungal spores based on measurements of three flow cytometry parameters (forward light scatter, wide-angle light scatter and DNA fluorescence) high percentage successful identification of *Fuligo septica, Oudemansiella radicata, Megacollybia platyphylla* and *Tylopillus felleus* (78%, 94%, 86% and 96% respectively) was obtained with only two nodes (60 000 training pattern presentations, 50 patterns per species; Morris et al. 1992). However, the mean output value of successfully identified patterns was low for one species, being 0.42, 0.83, 0.80 and 0.74 respectively for *F. septica*, *O. radicata*, *M. platyphylla* and *T. felleus*. With four hidden nodes, percentage successful identification altered little but respective mean output values were 0.69, 0.85, 0.90 and 0.80. Neither percentage successful identification nor mean output values improved significantly with further increase in number of hidden nodes, hence four-hidden-node networks were considered suitable in this particular case. The rule of thumb, mentioned above, would estimate the required number of hidden nodes for three inputs and four outputs as five.

Frankel et al. (1989) in their phytoplankton study with five input parameters (forward light scatter, fluorescence in the red (660–700 nm) and orange-red (540–630 nm) excited by each of two laser lines) and five outputs (two different sizes of beads, a group of various types of *Synechococcus*, a group of large phytoplankton and a group of prochlorophytes) achieved satisfactory learning with six and eight nodes in one hidden layer, but not with three hidden nodes.

Successful identification increases with increasing number of presentations of the training data (i.e. cycles or epochs) up to a point after which further presentations effect no improvement. In fact, neural networks can sometimes "over-train" if they are trained for too many cycles, that is they become too specific for the training set and their performance is poor with test data (Hecht-Nielsen 1990). In the fungal spore study (Morris et al. 1992), with low numbers of training cycles learning was incomplete, as evidenced by the low percentage of correct identifications of *F. septica* at or below 25 cycles (=5000 pattern presentations) and of *M. platyphylla* with 12.5 cycles (=2500 pattern presentations) for the four-hidden-node network (Fig. 12.5a). Again, while percentage identification of some species was good after relatively few cycles, the mean output node value of an identified species was low (Fig. 12.5b). Thus, it was considered that at least 200 cycles should probably be employed. There was no evidence that over-training occurred in this study. In the Frankel et al. (1989) phytoplankton study, their eight-hidden-node network learned to discriminate between the five types of cell after 80 cycles through a training set of 4800 cells.

With regard to the number of different training patterns necessary for successful learning, in the fungus spore study a four-hidden-node network was fairly well trained with just ten different patterns per species. The patterns used must have been representative of the whole sample, and a larger number of different training patterns would probably be necessary when there is more variation in data sets.

The Potential of Neural Computing Techniques for Analysis of Flow Cytometry Data

Clearly neural computing networks have been successfully applied to a few fairly trivial flow cytometry identification problems, and to much more complex non-biological pattern recognition problems (e.g. Caudill and Butler 1987). However, they should not be seen as the panacea for all

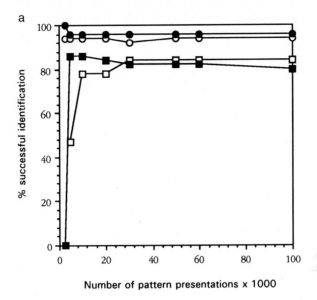

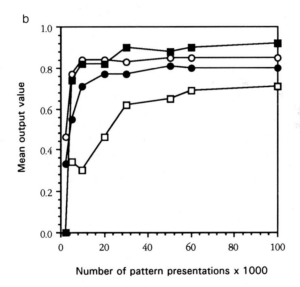

Fig. 12.5a,b. Effect of number of training cycles on learning four fungal spore types. **a** Effect on percentage successful identification. **b** Effect on mean output node value of successfully identified species. □, *F. septica*; ○, *O. radicata*; ■, *M. platyphylla*; ●, *T. felleus*. (Data from Morris et al. (1992).)

identification problems, since successful training is dependent upon measured parameters being good discriminators. In particular, if there is overlap in the distribution of input patterns for different species, the percentage successful identification will decrease. If several species have almost identical distribution of input patterns, then discrimination between these cannot be achieved using those data. However, by measuring additional parameters discrimination between such species should be possible. Since neural networks can cope with a huge number of input nodes this provides wide scope, but since most commercially available flow cytometers can usually only measure up to eight parameters simultaneously, discriminatory parameters must be chosen carefully. The number of measured variables can be increased by customizing machines. Alternatively, or additionally, partial discrimination between cell types could be achieved by neural network analysis of data obtained from a first run through the flow cytometer, and the broad categories of different cells collected by a cell sorter. Better discrimination could be achieved by making further flow cytometric measurements on the separate broad categories of cells, and subjecting the data to further neural network analysis. The greater the number of species used in training the greater the likelihood that the distribution of some of the input patterns will overlap or be almost identical, necessitating the use of more input parameters. This has proved to be the case when trying to distinguish between fifteen species of pathogenic bacteria (C.W. Morris, L. Boddy and R. Allman, unpublished data). *Klebsiella pneumoniae*, *Proteus mirabilis*, *Salmonella typhymurium* and *Staphylococcus aureus* could not be discriminated with only forward light scatter, wide-angle light scatter and DNA fluorescence as network input parameters; neither could *Legionella pneumophila* be discriminated from *Listeria monocytogenes*. However, the use of additional parameters looks more promising.

Another problem when discrimination between many different species is required is that a large number of different gain settings would probably be employed. Converting values from vastly different gain settings to the same scale might result in lack of "resolution", as the corrected values have a small numerical difference between them. This difficulty could be overcome by employing different neural networks for different gain settings. Another possibility might be to use gains as additional input parameters.

Neural network analysis of flow cytometry data can provide estimates of the species composition of mixed populations of cells, for species with which the network has been trained. However, it must be realized that the network will only be able to identify a species if it has already been trained upon appropriate input patterns for that species. If a network were presented with test data on a species upon which it had not been trained, an output code for a species upon which it *had* been trained would be generated. If all of the values in the output code were low, this might indicate that the identity of the cell was unknown, hence other identification methods would need to be employed, although low output code values also result when measured parameters of two known species overlap. If measured parameters of an unknown cell are identical with those of a species upon which the network was trained, there will be no indication that the identity of the cell type is unknown. The problem of "unknowns" may be particularly evident when identification of species from mixed field populations is attempted using neural networks which have been trained on data from laboratory pure cultures. Field populations of microorgansims may have more variable characteristics than those grown under constant, optimum conditions, and it will be essential to cover this variation in the training data set.

In conclusion, neural computing networks clearly have great potential for identification of cells from flow cytometry data. There is even the possibility of real-time, on-line data analysis since, while training networks can take many hours, identification of a single cell takes milliseconds, i.e. a comparable rate to generation of output from the flow cytometer. Much research, however, remains to be done before this potential can be realized.

References

Anon (1991) Neural Windows. Ward Systems Group Inc., Maryland, USA

Balfoort HW, Snoek J, Smits JRM, Breedveld LW, Hofstraat JW, Ringelberg J (1992) Automatic identification of algae: neural network analysis of flow cytometric data. J Plankton Res 14:575–589

Boddy L, Morris CW, Wimpenny JWT (1990) Introduction to neural networks. Binary 2:179–185

Caudill M, Butler C (eds) (1987) IEEE First international conference on neural networks, vol. 4. IEEE, San Diego

Dayhoff JE (1990) Neural network architectures: an introduction. Van Nostrand Reinhold, New York

Forsyth R (1990) Neural learning algorithms: some empirical trials. Proceedings of the third international workshop on neural networks applications, Nimes

Frankel DS, Olson RJ, Frankel SL, Chisolm SW (1989) Use of a neural net computer system for analysis of flow cytometric data of phytoplankton populations. Cytometry 10:540–550

Hecht-Nielsen R (1990) Neurocomputing. Addison-Wesley, New York

McClelland JL, Rumelhart DE (1988) Explorations in parallel distibuted processing: a handbook of models. MIT Press, Cambridge, Mass

McClelland JL, Rumelhart DE, PDP Research Group (1986) Parallel distibuted processing: explorations in the microstructure of cognition, vol. 2. Psychological and biological models. MIT Press, Cambridge, Mass

Morris CW, Boddy L, Allman R (1992) Identification of basidiomycete spores by neural network analysis of flow cytometry data. Mycol Res 96:697–701

NeuralWare Inc (1991) Reference guide NeuralWorks professional II/PLUS and NeuralWorks Explorer. NeuralWare Inc., Pittsburgh, USA

Noehte S, Manner R, Hausmann M, Horner H, Cremer C (1989) Classification of normal and aberrant chromosomes by an optical neural network in flow cytometry. Optical computing 1989 technical digest series. Optical Society of America, Washington, DC, pp 14–17

Rumelhart DE, McClelland JL, PDP Research Group (1986) Parallel distibuted processing: explorations in the microstructure of cognition, vol. 1. Foundations. MIT Press, Cambridge, Mass

Schwartz TJ (1990) A neural chips survey. AI Expert 5(12):34–38

Chapter 13

Rapid Analysis of Microorganisms Using Flow Cytometry

Michael Brailsford and Stephen Gatley

Introduction

A consistent limitation to the implementation of proactive product quality management has been the delay associated with microbiological analysis. Although standard microbiological techniques allow the detection of single bacteria, amplification of the signal is required through growth of a single cell into a colony on a plate. This process can be relatively time-consuming. For example, where an organism has a mean doubling time of approximately 30 min, the development of a colony containing 10^6 organisms (i.e. visible to the naked eye) will take between 18 and 24 h. In the case of yeast, a common spoilage organism, this period will be considerably longer, taking 3–7 days depending upon the yeast strain. In the case of final product testing, especially for products with a short shelf-life, microbiological data may not therefore be available until several days after the product has been released to the market. This can lead to product recall, with associated cost implications and reduction in customer confidence.

The emergence of rapid methods for the detection of microorganisms has largely been stimulated by the extended detection time of classical microbiological techniques, and has centred around the generation of alternative signals to reduce the time taken to obtain a detection. The parameters chosen for detection have generally been of a global nature – for example changes of electrical conductance or increases in turbidity (Boling et al. 1973; Sharpe et al. 1970; Cady 1975; Easter and Gibson 1989). These global signals are limited in the amount of information they can provide and are dependent on the fidelity of the signal produced. These limitations of existing rapid microbiology systems have restricted their integration into routine laboratory use. The development of systems which overcome these limitations has depended on the development of instruments which can "resolve" the analysis into cell-sized "bits" of information (Gatley 1990). In other words, the signal given by individual cells must be detectable by the analyser. This requirement has restricted the choice of technologies available to achieve this goal, with optical analysis being a leading contender.

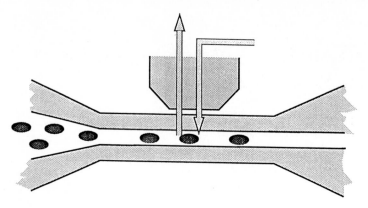

Fig. 13.1. Flow cytometry: automated single cell detection using optics and fluorescent markers.

In its simplest form, optical analysis can be undertaken using a microscope. However, microscopy has the disadvantage of requiring skilled operators and can be extremely time-consuming. A new generation of automated optical analysers and cell labelling technologies are now available, which incorporate the benefits of cell-by-cell analysis whilst overcoming the disadvantages of classical microscopy. These analysers are based on the principle of flow cytometry and provide rapid, semi-automated analysis of microorganisms.

Flow Cytometry

Flow cytometry may be considered as a form of automated fluorescence microscopy in which, instead of a sample being fixed to a slide, it is injected into a fluid which passes under the objective via a hydrodynamic focussing flow cell (Fig. 13.1). In the flow cell, the sample passes through a beam of light which results in the labelled cells emitting fluorescent pulses, each pulse being detected by a photomultiplier tube. Computer analysis then allows the pulses to be registered as separate counts and graded in terms of their relative fluorescence intensities.

The application of flow cytometry to industrial environments has required the development of a robust instrument designed specifically for routine use (Gatley 1989). The flow cytometry system used in the studies described below (the ChemFlow AutoSystem II) is capable of detecting up to 400 cells per second and the process of labelling and counting of cells can be carried out in under 30 min. This analyser has been developed for routine use and has now been evaluated in a range of industrial quality assurance laboratories (Desnouveaux et al. 1990; Bankes et al. 1991; Chevillotte et al. 1990).

The data presented in the "Applications" section below have been generated using fluorescent viability labels (Chemunex SA), a brief introduction to which follows.

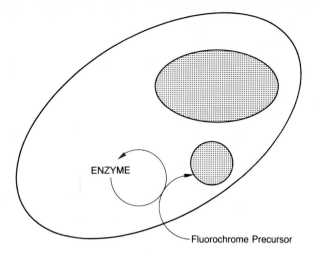

ENZYME

Fluorochrome Precursor

Fig. 13.2. Viability labels.

Viability Labels

Viable counts are performed using substrate molecules which, after entry into the cell, are cleaved by cytoplasmic enzymes liberating free fluoro-chrome into the internal matrix of the cell (Fig. 13.2). The intensity of the resulting intracellular cytoplasmic fluorescence is linked to the metabolic activity of the cell, with non-viable cells remaining unlabelled. This fluorescent signal is detected by the analyser, which thus differentiates live from dead cells.

Table 13.1. Analysis of yeast contamination in fruit products

Yeast inoculation level (CFU/g)	Pre-incubation time			
	24 h		48 h	
	Petri (+5 days)	ChemFlow	Petri (+5 days)	ChemFlow
S. cerevisiae				
0	−	−	−	−
0.7	+	+	+	+
6	+	+	+	+
39	+	+	+	+
Z. bailii				
0	−	−	−	−
1	+	−	+	+
9	+	+	+	+
70	+	+	+	+

Table 13.2. Detection of yeast in fromage frais

Inoculation level (yeast/g)	ChemFlow result after 18 h pre-incubation
0	−
0.1	+
1	+
10	+
100	+

Applications of Flow Cytometry in Industrial Quality Assurance

Analysis of Fruit Preparations

Determination of yeast and mould contamination is a key issue for producers and users of fruit products. Rapid detection of contamination of fruit products has been the subject of a number of recent studies with the ChemFlow system. The data in Table 13.1 show that the analyser is capable of detecting yeast in fruit products at levels of above 1 yeast/10 g product in 24–48 h. It is possible, therefore, to provide rapid positive release for fruit tanks prior to their use in yoghurt manufacture. This reduces the need for large cold-store facilities for tanks awaiting positive release from plate counts and improves the efficiency of fruit product manufacture and dispatch.

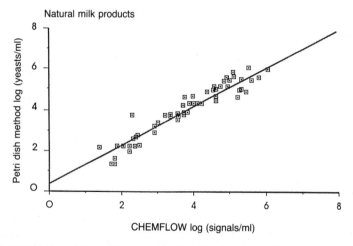

Fig. 13.3. Correlation of ChemFlow results with the plate count (Petri dish) method. The equation for the regression line is $y = 0.363 + 0.969x$, $r = 0.96$.

Table 13.3. Correlation between ChemFlow, standard plate count and shelf life

ChemFlow Signals/g product	Plate count CFU/g product	Shelf-life Expiry date (10 °C)
Samples between 100 and 2000	Positive from 100 to 5.2×10^3	Products acceptable
Samples >2000	Positive >2.5×10^3	Products not acceptable

Table 13.4. Relative incubation times required for the detection of high- and low-level yeast contamination in fermented milk products

Inoculation level (yeast/g)	Product type	Incubation times (h) required for detection
1	Natural	18
10	Natural	18
100	Natural	3
1000	Natural	2
1	Fruit-based	18
10	Fruit-based	6
100	Fruit-based	2
1000	Fruit-based	1

Analysis of Milk Products

Table 13.2 shows detection of yeast in fermented milk products. Levels as low as 0.1 yeast per gram product were detected in less than 24 h. Good correlation was observed with the standard plate count method (Fig. 13.3).

The data could also be correlated with the shelf-life of products (Table 13.3). It can be seen that the cut-off levels correlated well with subsequent product quality. At a viable count of <2000 counts per gram, product at the end of its shelf life was within specification. Above 2000 counts per gram all products were out of specification at the end of their shelf-life. It should be noted that a similar correlation could not be achieved using the standard plate count, where counts of up to 5.2×10^3 CFU per gram were obtained for both "in-spec" and "out-of-spec" products.

Through the determination of appropriate cut-off levels it is therefore possible to predict product quality at the end of its shelf-life prior to shipment to the customer.

Process Control

In addition to final product testing the ChemFlow has been evaluated for use as a process monitoring tool. Results presented in Table 13.4 show the levels of contamination compared with the time of detection. In both natural and fruit-based product types, inoculation of 100–1000 yeasts per gram could be detected within 3 h of product sampling. The slight differences

Table 13.5. Analysis of yeast contamination in salad products: spoilage index determination

Product type	Spoilage index	
	7 h at 28–30 °C	41 h at 4 °C
Coleslaw/natural	9.9	8.4
Coleslaw/fruit	0.8	1.0
Potato salad	0.5	0.5

between the two experiments reflects the yeast lag phase difference in the two product types. This approach can be used as a process line monitoring system or to allow rapid positive release of constituent products, e.g. white base prior to fruit addition. Furthermore it allows rapid intervention when a problem arises, thus minimizing plant down-time and production losses.

Shelf-Life Prediction in Salads and Fruit Juice Manufacture

In certain products, for example vegetable salads and fresh fruit juices, total numbers of viable yeast may be relatively high. In such cases their ability to grow in and spoil the product is the major factor in shelf-life determination. ChemFlow has been evaluated as a predictive tool for determining product shelf-life. Table 13.5 shows yeast counts determined in products before and after incubation of samples for 7 h at 30 °C. Samples were also stored at 4 °C for 41 h and tested at the end of this period to determine the natural growth of yeast at normal storage temperatures. The results were used to calculate a

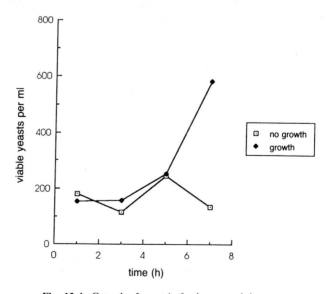

Fig. 13.4. Growth of yeast in fresh orange juice.

spoilage index (count after incubation expressed as a factor of initial count). Determination of spoilage at elevated temperature showed good correlation with the spoilage of the product at normal storage temperatures, and provided good predictive data on product quality.

Studies with fresh fruit juices (Fig. 13.4) indicated that total yeast counts at filling were not necessarily predictive of product shelf-life. High initial yeast counts did not necessarily correlate with subsequent product spoilage. The increase in counts following pre-incubation (i.e. spoilage index) provided a more accurate indication of product quality.

Fermenter Biomass and Spore Viability Analysis

Studies to evaluate the use of flow cytometry in fermentation monitoring have demonstrated a good correlation between the plate count and the ChemFlow counts for both ungerminated and germinated spores of

Table 13.6. Analysis of viable spores: correlation between the plate count and the ChemFlow count

Spore treatment	*Streptomyces* spores		*Penicillium* spores	
	ChemFlow (count/ml)	Plate (count/ml)	ChemFlow (count/ml)	Plate (count/ml)
Control	0	ND[a]	0	ND
Ungerminated[b]	3.6×10^7	5.8×10^7	8.2×10^6	7.3×10^6
Germinated 12 h	ND	ND	1.8×10^6	7.3×10^6
Ungerminated/UV[c]	2.8×10^7	8.5×10^4	5.4×10^6	8.8×10^4
Germinated 12 h/UV	ND	ND	1.7×10^5	8.8×10^4

[a] ND, not done; [b] ungerminated, spores stored at 4 °C; [c] UV, UV-treated spores.

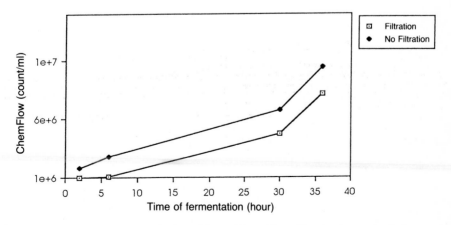

Fig. 13.5. Biomass monitoring with or without 15 mm filtration (no sonication).

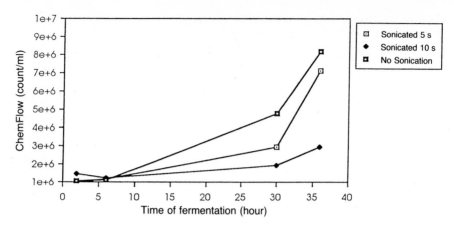

Fig. 13.6. Biomass monitoring with or without sonication (no filtration).

Penicillium and *Streptomyces* (Table 13.6). For spores treated with ultra-violet light (UV) and subsequently stored at 4 °C to inhibit cell growth (i.e. ungerminated spores) there was a poor correlation between the plate count and the ChemFlow count. Germination of UV-treated spores for 12 h prior to ChemFlow analysis restored the correlation between the plate count and the ChemFlow count.

These data may be explained by considering the mechanism of cell damage after UV treatment. This treatment results in a dysfunction of the cell repair mechanism in the affected cells that leads to cell damage during subsequent cell growth. Cell damage is therefore only apparent after a period of cell growth. The data presented here suggests that an incubation of approximately 12 h is sufficient to allow an accurate estimation of total viable spore count after UV treatment of a cell population.

In addition to the analysis of ungerminated spores, the development of biomass over the initial 36 h of a *Streptomyces* fermentation was studied. An increase in biomass was shown in both filtered and unfiltered samples (Fig. 13.5). The discrepancy between the two may indicate some loss of viable cells on the filter surface.

The effects of sonication on total cell counts is shown in Fig. 13.6. It can be seen that in all cases there was an increase in biomass. However, sonication decreased total counts relative to unsonicated samples, for both the 30 h and 36 h samples. This indicates that both fragmentation and cellular disruption may be occurring during sonication. No significant difference was observed between the sonicated and unsonicated samples at 2 and 6 h. This may be a function of the lower mycelial development in these samples.

These data indicate the potential of flow cytometry for "real-time" evaluation of spore viability and fermentation processes. The development of a simple, robust flow cytometer has opened new opportunities for this technology in the industrial sector.

Conclusion

Flow cytometry has been shown to be suitable for use throughout the manufacturing process and provides rapid data to aid the management of fermentation processes, processing plant hygiene, and raw material and final product quality. This allows the quality assurance manager to predict product quality throughout the production process, leading to reduced product recall, lower product losses and, ultimately, more cost-effective operations.

For juice and salads manufacturers the ChemFlow system can provide a rapid determination of total counts and also the capability of generating more predictive data on the growth of yeast in the product and thus a prediction of shelf-life. This allows more careful control of production through raw material and process line testing and allows positive release of high-quality product to the customer.

Future use of the system to detect microorganisms through the use of specific imunochemical labels such as monoclonal antibodies will ensure that this new approach to rapid microbiological analysis provides new solutions to an even wider range of industrial users.

Acknowledgements

The authors would like to acknowledge the contribution of Chemunex's Research and Development team (Director, Jean Louis Drocourt) who have provided extensive data and advice. Acknowledgement is also made to industrial co-workers, in particular Union Laiterie Normande, Grand Metropolitan Foods, and SmithKline Beecham. The authors would also like to thank the University of Wageningen (Holland) for their data on the fruit studies.

References

Bankes P, Rowe D, Betts RP (Campden Food and Drink Research Association, UK) (1991) The rapid detection of yeast spoilage using the ChemFlow System. Technical memorandum no. 621, May 1991

Boling EA, Blanchard GC, Russel WJ (1973) Bacterial identifications by microcalorimetry. Nature 241:472–473

Cady P (1975) Rapid automated bacterial identification by impedance measurements. In: Heden CG, Illeni T (eds) New approaches to the identification of micro-organisms. Wiley, Chichester, pp 74–99

Chevilbtte L, Laplace-Builhe C, Louvel L, Van Hoegaerden M, Theilleux J (1990) Application of flow cytometry to industrial microbiology. In: 5th European Congress on Biotechnology, Copenhagen, p 304 (abstr)

Desnouveaux R, Lecomte C, De Colombel E et al. (1990) Use of flow cytometry for yeast and mould detection in process control of fermented milk products: the ChemFlow system. A factory study. Biotech Forum Europe 3:224–229

Easter MC, Gibson DM (1989) Detection of micro-organisms by electrical methods. Progr Ind Microbiol 26:57–199

Gatley S (1989) Rapid process control with flow cytometry. Food Technology International-Europe

Gatley S (1990) Digital microbiology: a radical approach to the design and development of a new rapid microbiology system. Biotech Forum Europe 6:478–482

Sharpe AN, Woodron HN, Jackson AK (1970) ATP levels in food contaminated by bacteria. J Appl Bacteriol 33:758–767

Chapter 14

Appendix: Spectral Characteristics of Some Fluorescent Dyes and Excitation Sources

David Lloyd

In this table excitation (¦) and emission (|) maxima are indicated for a range of fluorochromes.

(Table overleaf)

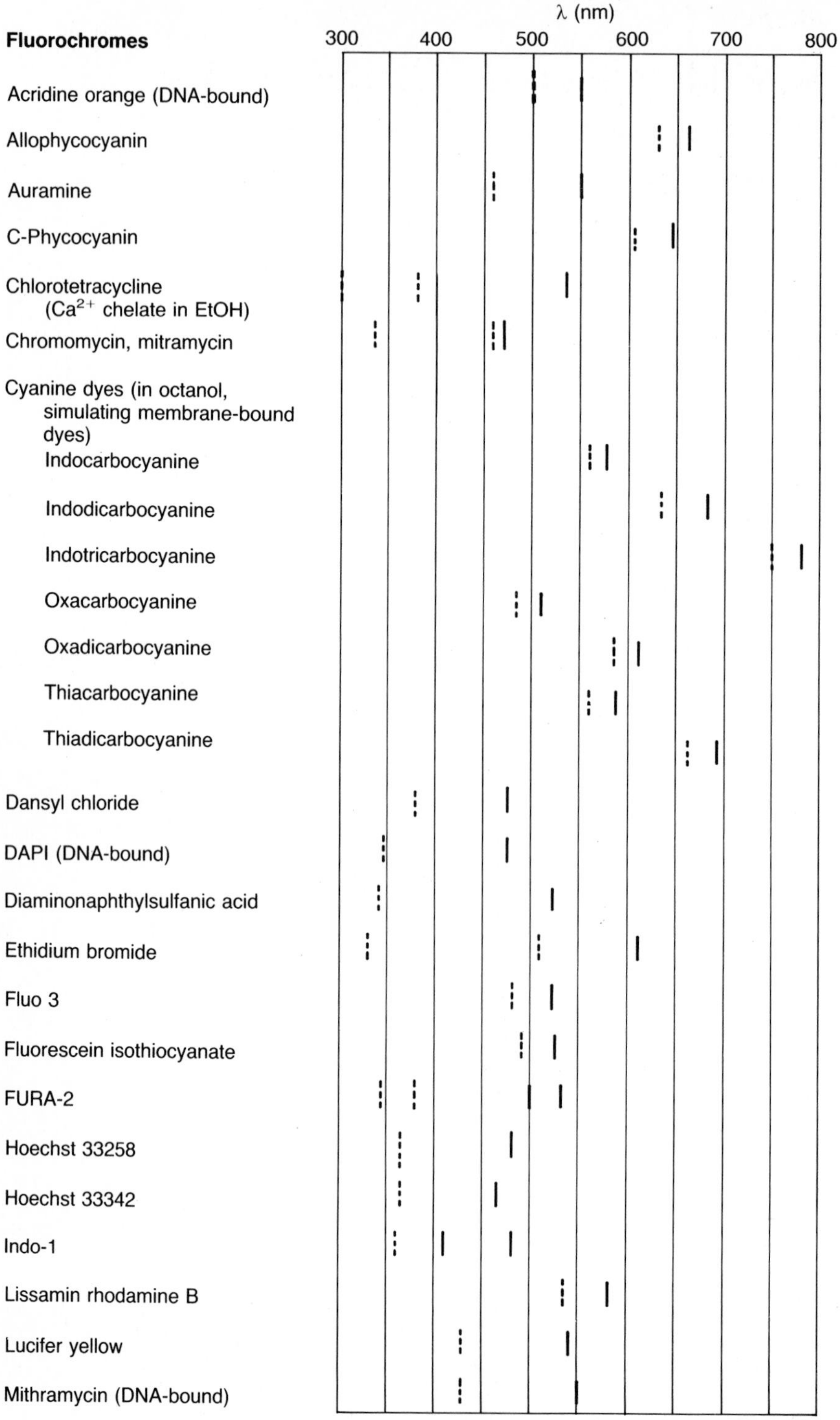

λ (nm)

Fluorochromes

300 400 500 600 700 800

Acridine orange (DNA-bound)

Allophycocyanin

Auramine

C-Phycocyanin

Chlorotetracycline
(Ca^{2+} chelate in EtOH)

Chromomycin, mitramycin

Cyanine dyes (in octanol,
simulating membrane-bound
dyes)
Indocarbocyanine

Indodicarbocyanine

Indotricarbocyanine

Oxacarbocyanine

Oxadicarbocyanine

Thiacarbocyanine

Thiadicarbocyanine

Dansyl chloride

DAPI (DNA-bound)

Diaminonaphthylsulfanic acid

Ethidium bromide

Fluo 3

Fluorescein isothiocyanate

FURA-2

Hoechst 33258

Hoechst 33342

Indo-1

Lissamin rhodamine B

Lucifer yellow

Mithramycin (DNA-bound)

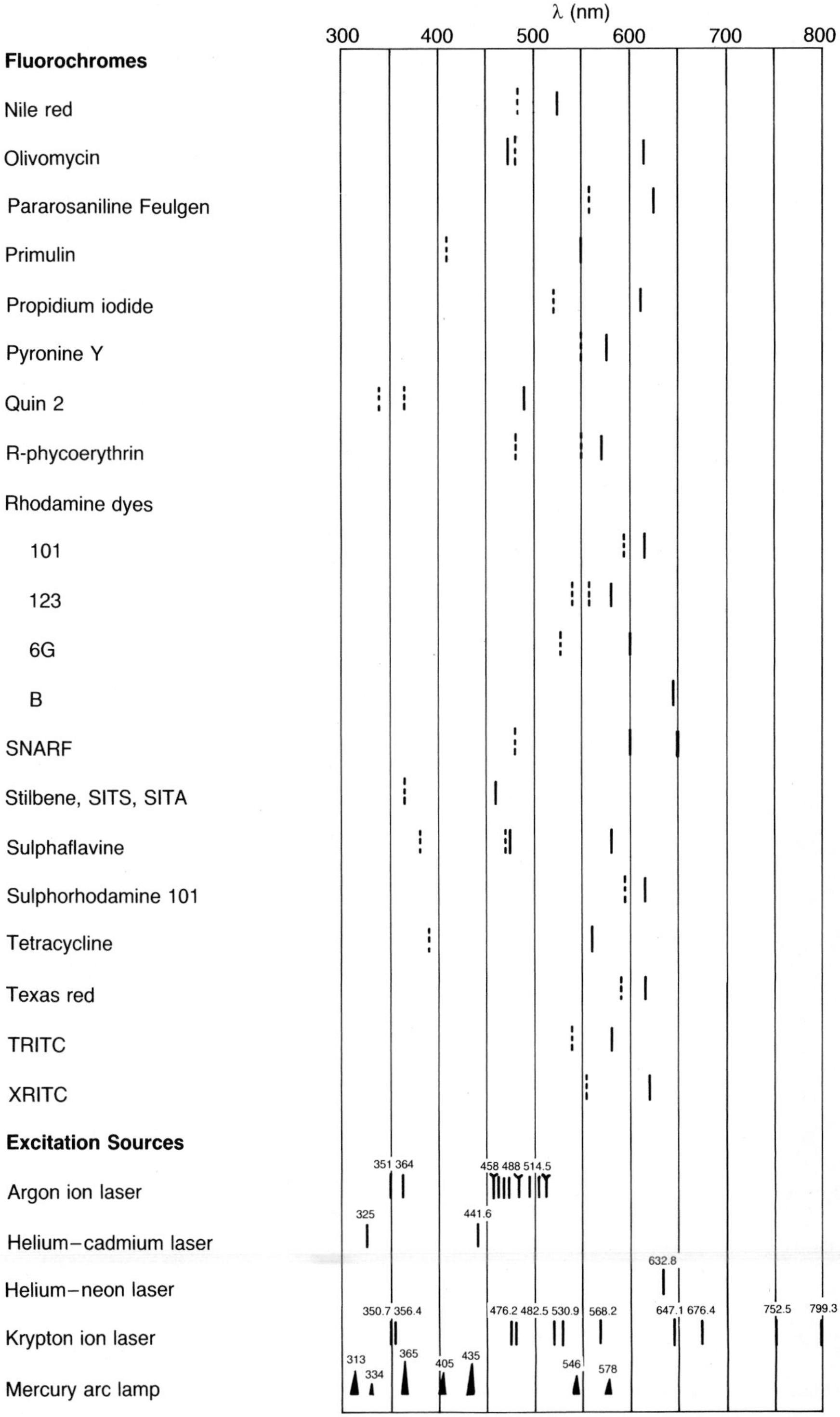

λ (nm)

300   400   500   600   700   800

Fluorochromes

Nile red

Olivomycin

Pararosaniline Feulgen

Primulin

Propidium iodide

Pyronine Y

Quin 2

R-phycoerythrin

Rhodamine dyes

101

123

6G

B

SNARF

Stilbene, SITS, SITA

Sulphaflavine

Sulphorhodamine 101

Tetracycline

Texas red

TRITC

XRITC

Excitation Sources

Argon ion laser   351 364   458 488 514.5

Helium–cadmium laser   325   441.6

Helium–neon laser   632.8

Krypton ion laser   350.7 356.4   476.2 482.5 530.9   568.2   647.1 676.4   752.5   799.3

Mercury arc lamp   313 334   365   405 435   546 578

Subject Index

Printing: Druckerei Zechner, Speyer
Binding: Buchbinderei Schäffer, Grünstadt

FI